Wetland Conservation and Ecological Restoration

Wetland Conservation and Ecological Restoration

Guest Editors

Qing Wang
Tian Xie
Jiakai Liu

MDPI

Basel • Beijing • Wuhan • Barcelona • Belgrade • Novi Sad • Cluj • Manchester

Guest Editors
Qing Wang
Research and Development Center for Watershed Environmental Eco-Engineering
Beijing Normal University
Beijing
China

Tian Xie
School of Environment
Beijing Normal University
Beijing
China

Jiakai Liu
School of Ecology and Nature Conservation
Beijing Forestry University
Beijing
China

Editorial Office
MDPI AG
Grosspeteranlage 5
4052 Basel, Switzerland

This is a reprint of the Special Issue, published open access by the journal *Water* (ISSN 2073-4441), freely accessible at: https://www.mdpi.com/journal/water/special_issues/OBBL1H8623.

For citation purposes, cite each article independently as indicated on the article page online and as indicated below:

Lastname, A.A.; Lastname, B.B. Article Title. *Journal Name* **Year**, *Volume Number*, Page Range.

ISBN 978-3-7258-6652-6 (Hbk)
ISBN 978-3-7258-6653-3 (PDF)
https://doi.org/10.3390/books978-3-7258-6653-3

Contents

About the Editors

Qing Wang

Qing Wang is an Associate Professor in the State Key Laboratory of Wetland Conservation and Restoration, Beijing Normal University. Her interests are in wetland ecological processes and restoration, as well as watershed eco-hydrology. She has presided over 10 projects, including the National Natural Science Foundation of China's Youth Project, the Guangdong Provincial Natural Science Foundation's General Project, and the Global Environment Facility (GEF) Projects, etc. She has published over 60 papers in journals such as *Journal of Applied Ecology*, *Ecological Indicators*, and *Journal of Environmental Management* and has served as a young editorial board member for *HydroResearch* and the *Journal of Environmental Science and Technology*, a guest editor for journals such as *Frontiers in Marine Science*, *Frontiers in Plant Science*, *Water*, and *Sustainability*, and a reviewer for many environmental science journals.

Tian Xie

Tian Xie is an Associate Professor in the State Key Laboratory of Wetland Conservation and Restoration, Beijing Normal University. He is dedicated to research on wetland ecological protection and restoration. He has led 7 scientific research projects, including the National Natural Science Foundation of China's general projects, youth projects, sub-projects of key research and development projects, etc. In the past five years, he has published over 60 papers in high-level journals such as *Global Change Biology*, *Communications Earth & Environment*, and *Journal of Applied Ecology*. He has served as the executive editor of *HydroResearch* and a guest editor for SCI journals such as *Frontiers in Marine Science*, *Water*, and *Sustainability*.

Jiakai Liu

Jiakai Liu is affiliated with the School of Ecology and Nature Conservation, Beijing Forestry University, where he serves as an associate professor. His core research interests focus on terrestrial ecosystem restoration, biodiversity conservation in fragmented landscapes, and the ecological responses of wetland communities. He has led/participated in multiple key research projects, including a Youth Project supported by the National Natural Science Foundation of China (NSFC). His scholarly findings have been published in more than 50 papers. Additionally, he acts as a peer reviewer for several leading ecological publications and collaborates with cross-disciplinary teams to integrate theoretical ecological insights with practical conservation management, striving to improve the efficacy of regional biodiversity protection and ecosystem restoration initiatives.

water

MDPI

Editorial

Wetland Conservation and Ecological Restoration

Qing Wang [1,2,3,*], Tian Xie [1,3] and Jiakai Liu [4,5]

1 State Key Laboratory of Wetland Conservation and Restoration, Beijing Normal University, Beijing 100875, China; tianxie@bnu.edu.cn
2 Research and Development Center for Watershed Environmental Eco-Engineering, Advanced Institute of Natural Sciences, Beijing Normal University, Zhuhai 519087, China
3 Yellow River Estuary Wetland Ecosystem Observation and Research Station, Ministry of Education, Dongying 257200, China
4 School of Ecology and Nature Conservation, Beijing Forest University, Beijing 100083, China; jkliu@bjfu.edu.cn
5 Key Laboratory of Ecosystem Protection and Restoration in the Yellow River Basin, National Forestry and Grassland Administration, Beijing 100083, China
* Correspondence: wq@bnu.edu.cn

Wetlands, also known as the "kidney of the earth," are one of Earth's three major ecosystems, along with oceans and forests. They are widespread worldwide and provide valuable ecosystem services for humans [1]; however, they have also suffered large-scale deterioration and loss, and such events highlight their value and importance [2,3]. In the 21st century, wetland-related issues have alarmed the global community, manifesting through phenomena such as water quality deterioration, eutrophication, water body depletion, and loss of biodiversity [2,4].

To reverse the trend of wetland loss and degradation and mitigate negative impacts, a growing number of wetland restoration efforts have been undertaken in recent years. Ecologists, biologists, and environmentalists have been working to find more effective solutions to restore degraded wetland ecosystems on a global scale [5–9]. However, the efficiency and effectiveness of these methods vary according to different measures, locations, wetland types, and the state of the wetland ecosystem [10]. Moreover, ecological restoration projects may have negative impacts [11–13]. Understanding the basic ecological process mechanisms and recognizing the interactions of different components in wetland ecosystems is crucial for effective and efficient wasteland conservation.

The concepts of "nature-based solutions," "adaptive management," and "ecological networks" seem to offer promising prospects and are currently being used to reframe wetland restoration in areas such as critical uncertainties reduction, climate change adaptation, and mitigation strategies [6,9]. As we enter the United Nations' Decade of Ecosystem Restoration (2021–2030), countries and organizations worldwide will pay closer attention to the innovations underpinning ecological restoration to ensure that restoration efforts reach their full potential in delivering social and ecological coordination and, ultimately, sustainable development.

We aim to advance discussions on the various aspects of wetland conservation and ecological restoration, such as the key ecological processes of wetlands and their implications for restoration; nature-based solutions in wetlands; wetland ecological risks and management; wetland biodiversity conservation; wetland pollution and control; and wetland ecological function improvement. The seven articles and two review papers presented discuss biological aspects, biogenic element, carbon sink function and resilience in wetland, wetland ecological water replenishment, wetland conservation survey, ecological restoration suitability evaluation, and water purification in treated wetland.

 https://doi.org/10.3390/w17243484

Two of the presented articles discuss the biological aspects of natural wetlands. Chen et al. focus on wetland hydrology's influence on plants. They use hydrological monitoring data to study different flooding conditions and investigate the responses of a typical species in the lakeside wetland in Momoge National Natural Reserve at multiple levels, finding that maintaining seasonal flooding is essential for the natural restoration of *Bolboschoenus planiculmis* wetlands. These findings have important implications for the near-natural restoration of *Grus leucogeranus* habitats. Liu and Zhang consider the responses of microbial communities in urban wetland ecosystems to the atmospheric pollution, especially at three major interfaces of atmosphere, foliage and water. They describe how the structure and function of microbial communities (i.e., α- and β-diversity) change in response to different pollution levels, aiming to provide guidelines for monitoring urban wetland ecosystems and species diversity conservation.

Li et al. studied Dongting Lake, the second largest freshwater lake in China, which is connected to the Yangtze River and provides a valuable habitat for migratory birds. In their paper, they analyze the spatial distribution of nutrients (TOC, TN, and TP) in the soil along a hydrographic gradient and discuss the impact factors of the nutrient patterns. The results enhance our understanding of the relationship between soil nutrient variations and hydrological dynamics in the floodplain and have implications for the conservation of migratory bird habitats.

The two review papers pertain to wetland carbon sink function and resilience to drought. Guo et al. review the impact of ecological restoration on carbon sink function in coastal wetlands. They explain the concept of coastal wetland carbon sink function and its influencing factors; summarize various restoration projects, along with their objectives and measures at the national scale; and discuss the different impacts of restoration measures on the carbon sink function. Finally, they recommend integrating carbon sink function enhancement into the design and implementation of ecological restoration projects. This study provides a comprehensive understanding of the relationship between ecological restoration and carbon sink function, which has significant reference value and management implications for ecological restoration. Meanwhile, Lu et al. discuss plant drought resilience, which is an important indicator of wetland ecosystems' resilience to climate change. They review plant drought resilience at multiple scales, from molecular mechanisms to ecosystem sustainability, and predict future emphasis on multi-scale and multi-dimensional integrated analysis. Their work aims to provide a theoretical basis for ecosystem sustainability and agricultural production under climate change.

Meng et al. propose a novel framework for estimating wetland ecological water requirements. They conduct a case study in the Momoge Wetland, where the current water replenishment projects are inadequate to meet the ecological water demand. To promote the development of dynamic water replenishment strategies, especially during periods of drought, the authors highlight the urgent need to implement multi-source water replenishment techniques. Specifically, their study provides insights relevant to annual and seasonal water replenishment planning and multi-source water management of wetlands that are facing similar issues to the Momoge Wetland, which will help wetland ecosystems adapt to climate change in the future.

Mo and Yan's social science study provides suggestions for wetland conservation. They conducted a questionnaire-based survey in Beijing to systematically analyze the residents' awareness, attitudes, and behaviors regarding wetland conservation and provide constructive suggestions for urban wetland management practices. They point out that the geographical location and functional configuration of wetlands are the main factors influencing residents' behaviors, with higher visit rates recorded at wetlands closer to urban areas. Moreover, public awareness of wetlands' ecological functions is lacking, particularly with regard to biodiversity conservation and water quality improvement.

Miao et al. evaluate ecological restoration in an Agro-Pastoral Transition Zone and assess the importance of ecological functions, such as soil and water conservation, biodiversity maintenance, windbreak and sand fixation; and ecological sensitivity and protection. They also provide a scientific case study for local ecosystem restoration and conservation which will guide future ecological restoration efforts in such areas. In the future, multi-source data fusion, the establishment of a multi-scale evaluation system, and trade-off analysis between conservation and development will be explored in greater detail.

Moustafa et al. test the effects of physical parameters, such as flow, transport, and water depth, on phosphorus retention in treated wetlands, which could benefit the design of artificial wetlands by enhancing their purification efficiency. They suggest that optimizing flow and depth controls in wetland design and management is important for enhancing phosphorus removal efficiency in large constructed wetland systems.

Wetland conservation and ecological restoration are extensive topics with many unresolved questions. We anticipate that this compilation of papers will offer clear and valuable insights into wetland conservation and ecological restoration, presenting restoration, conservation, and management implications and specific strategies to address areas of uncertainty.

Author Contributions: Q.W.: Writing—review and editing, Writing—original draft. T.X.: Writing—review and editing. J.L.: Writing—review and editing. All authors have read and agreed to the published version of the manuscript.

Acknowledgments: The study was supported financially by the Natural Science Foundation of Guangdong Province (2025A1515011055) and the Natural Science Foundation of China (42107057). The editors are grateful to be contributions made by the many authors involved in this project.

Conflicts of Interest: The authors declare no conflicts of interest.

List of Contributions:

1. Chen, L.; Zhang, M.; Tong, S.; An, Y.; Zhao, C.; Xin, Y.; Zhang, J. Influence of Flood Events on the Ecological Characteristics of *Bolboschoenus planiculmis*: Implications for Restoration of Grus leucogeranus Habitats. *Water* **2024**, *16*, 3672. https://doi.org/10.3390/w16243672.
2. Li, J.; Wu, Y.; Peng, D.; Chen, M.; Peng, L.; Middleton, B.A.; Lei, T. Spatial Differences in Soil Nutrients Along a Hydrographic Gradient on Floodplains in Dongting Lake. *Water* **2024**, *16*, 3674. https://doi.org/10.3390/w16243674.
3. Liu, Y.; Zhang, Z. Effects of Atmospheric Particulate Matter on Microbial Communities in Wetland Ecosystems. *Water* **2025**, *17*, 66. https://doi.org/10.3390/w17010066.
4. Meng, H.; Zhong, X.; Wu, Y.; Peng, X.; Li, Z.; Wang, Z. Estimation of Ecological Water Requirement and Water Replenishment Regulation of the Momoge Wetland. *Water* **2025**, *17*, 114. https://doi.org/10.3390/w17010114.
5. Mo, L.; Yan, B. Awareness and Behaviors of Beijing Residents Regarding Wetland Conservation. *Water* **2025**, *17*, 375. https://doi.org/10.3390/w17030375.
6. Guo, X.; Liu, Y.; Xie, T.; Li, Y.; Liu, H.; Wang, Q. Impact of Ecological Restoration on Carbon Sink Function in Coastal Wetlands: A Review. *Water* **2025**, *17*, 488. https://doi.org/10.3390/w17040488.
7. Miao, J.-J.; Gao, Y.-H.; Zhang, Y.; Gao, X.-S.; Xu, D.-H.; Yang, J.-Q.; Wang, W.; Liu, H.-W. Suitability Evaluation of Ecological Restoration Relying on Water Resources in an Agro-Pastoral Transition Zone: A Case Study of Zhangbei, Zhangjiakou, Northern China. *Water* **2025**, *17*, 1393. https://doi.org/10.3390/w17091393.
8. Moustafa, M.Z.; Lal, W.A.M. Phosphorus Retention in Treatment Wetlands? A Field Experiment Approach: Part 2, Water Quality. *Water* **2025**, *17*, 1746. https://doi.org/10.3390/w17121746.

9. Lu, W.; Wu, B.; Wang, L.; Gao, Y. Multi-Scale Drought Resilience in Terrestrial Plants: From Molecular Mechanisms to Ecosystem Sustainability. *Water* **2025**, *17*, 2516. https://doi.org/10.3390/w17172516.

References

1. Barbier, E.B.; Hacker, S.D.; Kennedy, C.; Koch, E.W.; Stier, A.C.; Silliman, B.R. The value of estuarine and coastal ecosystem services. *Ecol. Monogr.* **2011**, *81*, 169–193. [CrossRef]
2. Newton, A.; Icely, J.; Cristina, S.; Perillo, G.M.E.; Turner, R.E.; Ashan, D.; Cragg, S.; Luo, Y.; Tu, C.; Li, Y.; et al. Anthropogenic, Direct Pressures on Coastal Wetlands. *Front. Ecol. Evol.* **2020**, *8*, 144. [CrossRef]
3. Murray, N.J.; Clemens, R.S.; Phinn, S.R.; Possingham, H.P.; Fuller, R.A. Tracking the rapid loss of tidal wetlands in the Yellow Sea. *Front. Ecol. Environ.* **2014**, *12*, 267–272. [CrossRef]
4. Ma, X.; Brookes, J.; Wang, X.; Han, Y.; Ma, J.; Li, G.; Chen, Q.; Zhou, S.; Qin, B. Water quality improvement and existing challenges in the Pearl River Basin, China. *J. Water Process. Eng.* **2023**, *55*, 104184. [CrossRef]
5. Li, S.; Li, Y.; Jiang, N.; Xu, W. Development of key ecological conservation and restoration projects in the past century. *Ecol. Front.* **2025**, *45*, 1–6. [CrossRef]
6. Cheng, C.; Li, F. Ecosystem restoration and management based on nature-based solutions in China: Research progress and representative practices. *Nat. Based Sol.* **2024**, *6*, 100176. [CrossRef]
7. Aradottir, A.L.; Hagen, D. Chapter Three—Ecological Restoration: Approaches and Impacts on Vegetation, Soils and Society. *Adv. Agron.* **2013**, *120*, 173–222.
8. Schuster, L.; Taillardat, P.; Macreadie, P.I.; Malerba, M.E. Freshwater wetland restoration and conservation are long-term natural climate solutions. *Sci. Total Environ.* **2024**, *922*, 171218. [CrossRef] [PubMed]
9. Ferreira, C.S.S.; Kasanin-Grubin, M.; Solomun, M.K.; Sushkova, S.; Minkina, T.; Zhao, W.; Kalantari, Z. Wetlands as nature-based solutions for water management in different environments. *Curr. Opin. Environ. Sci. Health* **2023**, *33*, 100476. [CrossRef]
10. Wang, X.; Xiao, X.; Xu, X.; Zou, Z.; Chen, B.; Qin, Y.; Zhang, X.; Dong, J.; Liu, D.; Pan, L.; et al. Rebound in China's coastal wetlands following conservation and restoration. *Nat. Sustain.* **2021**, *4*, 1076–1083. [CrossRef]
11. Wortley, L.; Hero, J.-M.; Howes, M. Evaluating Ecological Restoration Success: A Review of the Literature. *Restor. Ecol.* **2013**, *21*, 537–543. [CrossRef]
12. Sun, Z.; Sun, W.; Tong, C.; Zeng, C.; Yu, X.; Mou, X. China's coastal wetlands: Conservation history, implementation efforts, existing issues and strategies for future improvement. *Environ. Int.* **2015**, *79*, 25–41. [CrossRef] [PubMed]
13. Liu, N.; Ma, Z. Ecological restoration of coastal wetlands in China: Current status and suggestions. *Biol. Conserv.* **2024**, *291*, 110513. [CrossRef]

water

MDPI

Article

Influence of Flood Events on the Ecological Characteristics of *Bolboschoenus planiculmis*: Implications for Restoration of *Grus leucogeranus* Habitats

Long Chen [1], Mingye Zhang [1,*], Shouzheng Tong [1], Yu An [1], Chunzi Zhao [2], Yuan Xin [1] and Jiaxin Zhang [3]

1 State Key Laboratory of Black Soils Conservation and Utilization, Northeast Institute of Geography and Agroecology, Chinese Academy of Sciences, Changchun 130102, China; chenlong20230914@163.com (L.C.)
2 College of Geography and Ocean Sciences, Yanbian University, Hunchun 133300, China
3 School of Geographical Sciences, Changchun Normal University, Changchun 130032, China
* Correspondence: zhangmingye@iga.ac.cn

Abstract: Flood events severely damage the biodiversity and ecological functions of wetlands, posing a major threat to the health and stability of wetland ecosystems. Plants play a crucial role in maintaining the stability and balance of these ecosystems by providing food and habitat for various organisms. Although the wetland plants' responses to flooding events have been extensively studied, the multi-level ecological characteristics (on the community, population, and individual plant level) of these plants in response to flooding have not yet been investigated. In this study, the community structure and ecological characteristics of *Bolboschoenus planiculmis* under different flooding conditions and plant traits were studied. The results revealed significant differences in the community composition and species diversity under various flooding conditions. Under continuous flooding, the number of species was three times greater than under seasonal flooding conditions. Flood events showed a significant impact on population density and coverage of *B. planiculmis*. The population density and coverage were 76.10% and 66.70% higher in seasonal flooding conditions than in continuous flooding conditions. Under seasonal flooding conditions, the allocation of total biomass and bulb biomass was greater than that observed under continuous flooding conditions. The results of the correlation analysis and redundancy analysis (RDA) indicated that the water level is a critical factor influencing the variations in the multi-level ecological features of the *B. planiculmis* community under different flooding conditions. This study suggests that maintaining seasonal flooding is essential for the natural restoration of *B. planiculmis* wetlands. These findings demonstrate that flood events significantly affect the ecological characteristics of *B. planiculmis* and offer valuable guidelines for the near-natural restoration of *Grus leucogeranus* habitats.

Keywords: wetland restoration; near-natural restoration; multi-level ecological characteristics; water-level; hydrological control strategy

1. Introduction

Floods are among the most widespread and frequent natural hazards that affect surface ecosystems and cause significant destruction [1,2]. Globally, approximately 2.23 billion hectares of land are at risk of flooding, leading to significant alterations in surface ecology, including shifts in vegetation patterns, changes in soil composition, and disruption of local biodiversity [3,4]. Recent ecological studies indicate that fluctuations in hydrological conditions resulting from floods significantly affect the stability of net primary productivity in wetland ecosystems compared with other ecosystems [5–7]. These fluctuations pose a serious threat to the essential ecological services provided by wetland ecosystems, and may ultimately jeopardize human well-being [8]. Consequently, wetland hydrological management has become a priority for policymakers and wetland management organizations aiming to promote wetland conservation, particularly in floodplains where flooding events occur frequently [9,10]. Wetland plants, as integral components of these ecosystems, exhibit multilayered ecological organization [11,12]. These hierarchical arrangements, ranging from communities to populations to individual plants, collectively influence the biodiversity and ecological functions of wetlands [13,14]. Wetland biodiversity is most prominently reflected in plant species diversity, under typical conditions, in which a complex plant community composition positively contributes to the ecological functions of wetlands [15–17]. According to the bottom–up effect, a rich diversity of plant species can enhance the quality of biological habitats, thereby promoting long-term conservation and stabilization of populations of rare avian species [18–20]. The dynamics of wetland plant populations are examined by revealing the effects of environmental change on these plants through ecological characteristics, such as population density and cover [21]. Changes in hydrological conditions directly or indirectly influence community density and cover by affecting plant growth and development cycles as well as seed germination conditions [22–24]. Variations in plant growth strategies reflect how plants respond to environmental changes, including water regulation, nutrient uptake, growth rate regulation, and reproductive strategies, thereby highlighting the adaptive mechanisms plants use to overcome survival challenges. Specifically, the depth, frequency, and duration of flooding can directly affect biomass allocation and growth rate strategies in plants. Flooding creates submerged anaerobic environments where plant escape strategies enable branches, leaves, and adventitious root systems to emerge from anaerobic conditions. In contrast, the growth tolerance strategies of different plants directly influence their growth rate. Terrestrial plants that are intolerant to water may reduce or cease growth or perish under inundated conditions [25–27]. Previous studies have extensively investigated the responses of individual levels of ecological organization in plants to flood events; however, research analyzing the responses of multiple levels of ecological organization in tandem remains limited.

Bolboschoenus planiculmis is a perennial herbaceous plant belonging to the Cyperaceae family and is widely distributed across wetlands in East Asia, Central Asia, and Eastern Europe. The starch- and protein-rich bulbs of this species serve as vital food sources for whooping cranes and other rare birds [28,29]. However, extreme hydrological conditions resulting from climate change and human disturbances in recent decades have significantly reduced the distribution range of this species, leading to a shortage of food sources that severely reduces the ecological quality of its habitat [30,31]. The decline of *B. planiculmis* will disrupt the associated ecological chain and threaten the overall ecological security of the wetland. Therefore, scientific and effective restoration of *B. planiculmis* wetlands and their ecological functions are essential for maintaining the ecological security of rare bird populations [32]. Although previous studies have examined factors such as water and salinity, planting methods, and trigger induction, little is known about the effects of flooding events on the community and trait characteristics of *B. planiculmis* [28,33,34].

Located in the western part of the Songnen Plain, the Momoge Wetland serves as a crucial stopover for migratory birds along the East Asia–Australasia migratory route. The extensively developed wetlands of *B. planiculmis* in this region provide suitable habitats and ample food sources for white cranes [35–37]. However, the combined effects of climate change and anthropogenic disturbances have led to frequent flooding events, which have significantly affected the stability of *B. planiculmis* communities, resulting in habitat degradation and food shortages for rare avian species [30,38]. Although previous studies have examined individual *B. planiculmis* plants' responses to flooding events, the impact of inundation on multitiered levels of *B. planiculmis* wetlands, specifically from the perspective of community ecological characteristics and population traits, has not yet been reported. Therefore, to achieve near-natural restoration of *B. planiculmis*, this study aimed to (1) determine the ecological characteristics of plant communities under different flooding conditions, (2) investigate the responses of *B. planiculmis* population ecological traits to changes in flooding conditions, and (3) identify plant traits associated with growth under varying flooding conditions. We hypothesized that different flooding events would significantly affect the characteristics of the *B. planiculmis* communities and populations. The results of this study enhance our understanding of the responses of the *B. planiculmis* community and its trait characteristics to flooding events, providing valuable information for flood management in the habitats of rare avian species.

2. Materials and Methods

2.1. Study Area

The study area is located in the lakeside wetland of Ertou Lake (45°42′–46°18′ N, 123°27′–124°04′ E) (Figure 1), which constitutes the core region of the Momoge National Nature Reserve (MNNR). This region experiences an arid to semi-arid continental monsoon climate with an average annual precipitation of 380 mm, primarily occurring from June to August. Meanwhile, intense rainfall leads to flooding events that inundate the lakeside area. The average annual temperature is 4.4 °C, whereas annual evapotranspiration is approximately 1472 mm, and the area is characterized by meadow swampy soils. Typical native plant species found in the study area include *B. planiculmis*, *Phragmites australis*, and *Carex schmidtii* [39]. Climate change combined with agricultural drainage contributes to flood events in the MNNR, thereby affecting the normal growth of wetland plants [40,41]. Every year, 95% of the world's *Grus leucogeranus* migrate here to feed. Ertou Lake is an important habitat, with abundant *B. planiculmis* growing along the shore, providing a rich food source for the cranes.

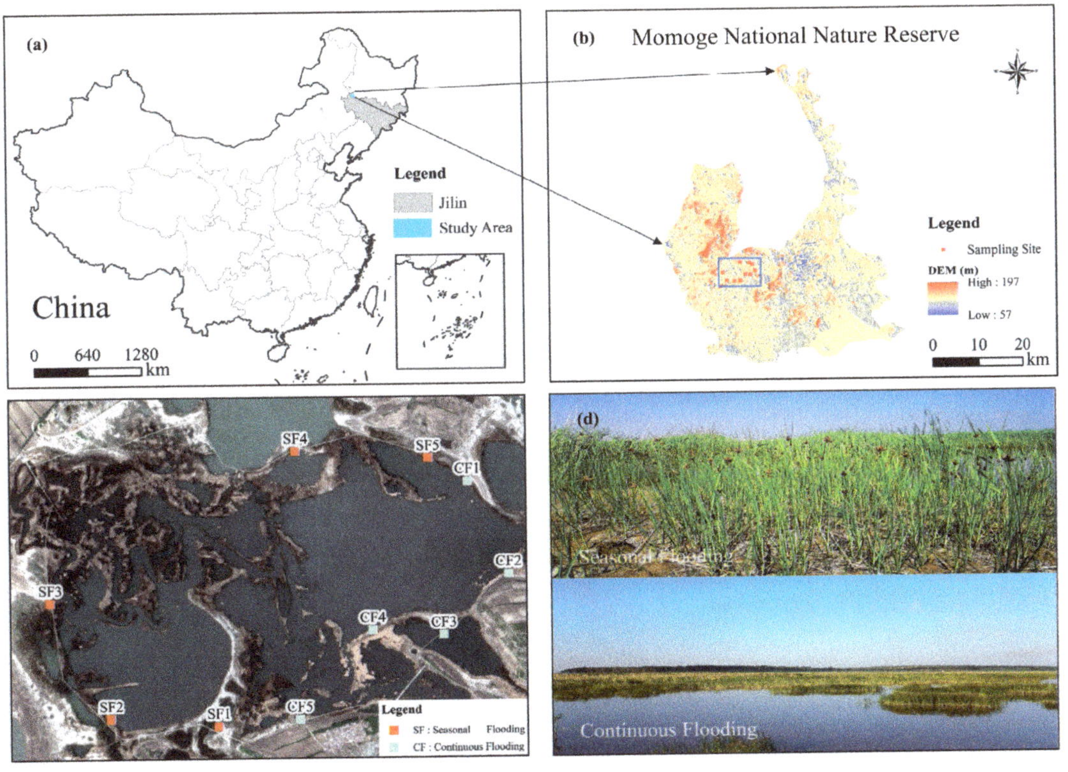

Figure 1. Map of the Momoge study area: (**a**) Momoge location map, (**b**) study area, (**c**) sampling sites, and (**d**) seasonal and continuous flooding areas.

2.2. Field Survey

In August 2019, a field survey was conducted at ten strategically selected sampling sites, based on prior hydrological monitoring data. These sites were chosen to represent distinct hydrological regimes: five seasonal flooding sites (SF1–SF5), characterized by periodic inundation lasting approximately five to six months annually, and five continuous flooding sites (CF1–CF5), which experienced year-round inundation. Each sampling site included a sample strip measuring 10 × 100 m. A 1 × 1 m sample plot was randomly established at 50 m intervals, resulting in a total of 30 sample plots. The plant species, density, height, and coverage were recorded for each plot. Plant species were identified according to the guidelines outlined in Chinese Vegetation [42]. Population density was determined using the counting method, height was measured using a steel ruler, and coverage was estimated using the vertical projection method. To monitor water-level fluctuations at the ten sampling sites, we deployed ten water-level recorders (Odyssey, Auckland, New Zealand) to collect the surface water-level data. The recording frequency was 3 times per day. The aboveground and belowground biomass of *B. planiculmis* was collected from the sample plots and transported to the laboratory in plant collection bags. Three soil samples were collected randomly from areas with similar background characteristics near the plant samples. Surface soil samples were collected from a depth of 0–20 cm and three samples were mixed to create one composite sample. The soil samples were sealed in bags and transported to the laboratory for analysis.

2.3. Sample Analyses

In this study, the species importance value (*Pi*) was used to evaluate the significance of each species within the community. Additionally, the species richness index (*R*), Shannon–Wiener diversity index (*W*), Simpson index (*N*), and Pielou evenness index (*P*) were utilized to characterize the species diversity features of the community, which were calculated according to the following equations:

$$P_i = \frac{(RH + RC + RD)}{3} \tag{1}$$

$$R = S \tag{2}$$

$$W = -\sum_{i=1}^{S} \ln P_i \tag{3}$$

$$N = 1 - \sum_{i=1}^{S} P_i^2 \tag{4}$$

$$P = \frac{W}{\ln S} \tag{5}$$

where *RH* represents the relative height, *RC* represents the relative coverage, *RD* represents the relative density, and S represents the total number of species.

The soil pH was measured using the electrode method described by Zhang et al. [43]. Electrical conductivity (EC) was determined using the electrode method, according to the procedure outlined by Qi et al. [44]. The moisture content (MC) was evaluated using the desiccation method, as reported by Yu et al. [45]. The bulk density (BD) was measured using the ring knife method, as described by Qi et al. [44], whereas soil particle density (PD) was determined using the specific gravity bottle method, as described by Yu et al. [45]. Soil porosity (SP) was calculated using the specified calculation method [45].

2.4. Data Analysis

An independent sample *t*-test was conducted using SPSS version 23 [46] to evaluate the differences in community species diversity, population ecological traits, and biomass under varying hydrological conditions ($p < 0.05$). Before the analysis, the data were assessed for normal distribution and homogeneity of variance. For the community diversity data and the importance values of *B. planiculmis*, which did not satisfy the assumptions of normality or homogeneity of variance after standardization, the Mann–Whitney test was used to analyze the differences between groups ($p < 0.05$). Correlation analyses were performed to examine the relationships between the diversity of *B. planiculmis* communities, soil environmental factors, ecological traits of the *B. planiculmis* population, and plant characteristics, utilizing the "linkET" package in R 4.3.0 [47]. Primary variables influencing the ecological traits of the *B. planiculmis* population were identified using Canoco version 5.0 [48], establishing a significance level for each explanatory factor. Based on the first axis length of Detrended Correspondence Analysis (0.88), redundancy analysis (RDA) was used to test the key soil environmental factors affecting the ecological traits of the population. Statistical analysis and visualization were performed using Excel 2021 [49] and Origin 2022 [50].

3. Results

3.1. Multi-Level Ecological Characteristics

3.1.1. Community Characteristics

A survey of the wetland plant community revealed the presence of six families, eight genera, and nine species within the community (Figure 2). Notably, three species were documented under SF conditions and nine were identified under CF conditions. The community exhibited a predominance of perennial plants, constituting 77.78% of the total, with only two species: Persicaria maculosa Gray and Utricularia aurea Lour were classified as annual, and both species were observed exclusively under CF conditions, perennials are more dominant than annuals in flood-prone environments. The composition of the plant community included 55.56% aquatic plants and 44.44% mesophytes, which is characteristic of typical lakeshore wetland plant communities. Although *B. planiculmis* was the dominant species across both hydrological conditions, its importance value was significantly lower under CF conditions, recorded at 0.97 ± 0.05 in SF conditions and 0.80 ± 0.10 in CF conditions ($p < 0.05$, Figure 3e). Furthermore, indices such as species richness, the Shannon–Wiener index, Simpson index, and Pielou index were all significantly higher under SF conditions than under CF conditions ($p < 0.05$, Figure 3b–d). These findings indicate that the diversity of *B. planiculmis* communities varied significantly owing to variations in the flooding conditions.

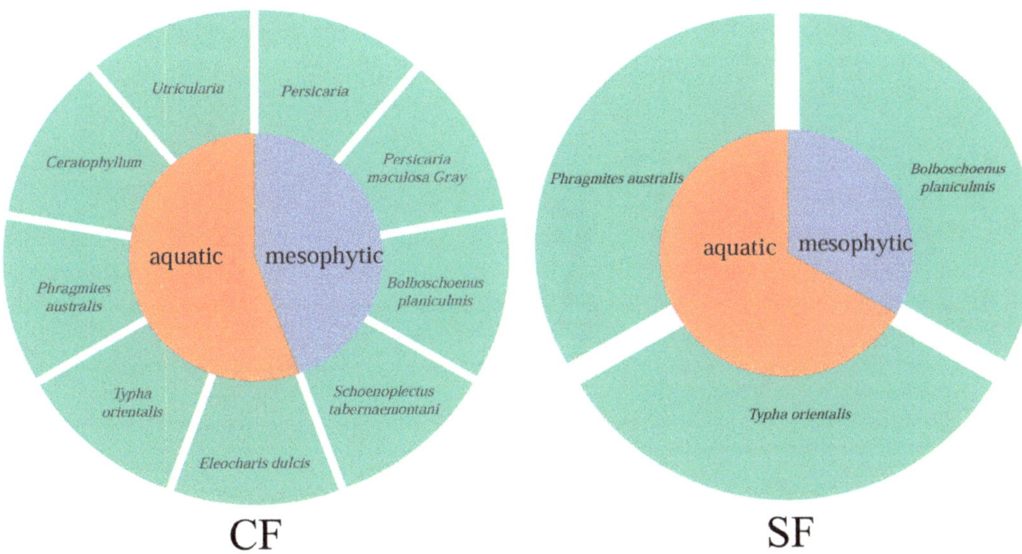

Figure 2. Species composition and growing environment of *Bolboschoenus planiculmis* communities under different flooding conditions: SF—seasonal flooding conditions; CF—continuous flooding conditions.

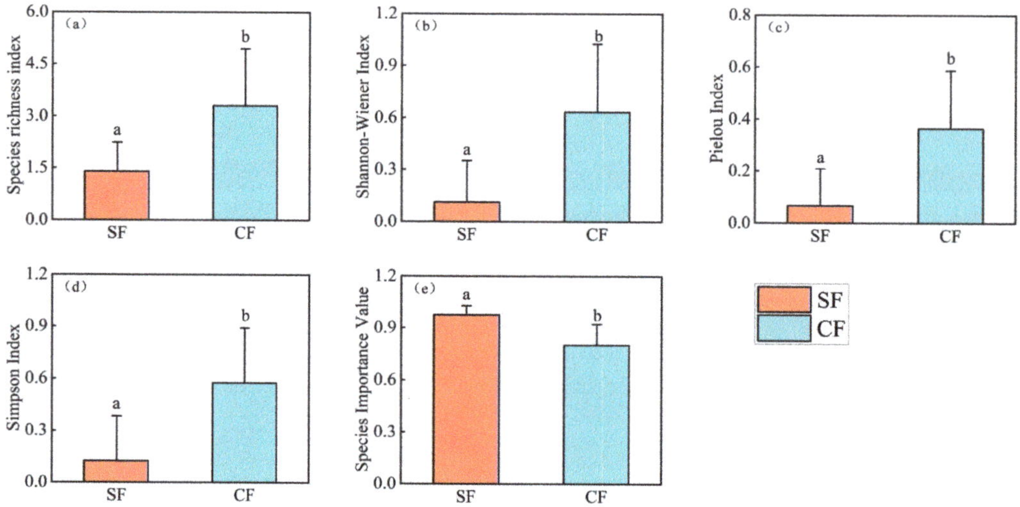

Figure 3. *Bolboschoenus planiculmis* community diversity and importance value under different flooding conditions (mean ± standard error, n = 15): SF—seasonal flooding condition; CF—continuous flooding condition; (**a**) species richness; (**b**) Shannon–Wiener index; (**c**) Pielou index; (**d**) Simpson's index; and (**e**) *Bolboschoenus planiculmis* importance value. Different letters represent significant differences ($p < 0.05$).

3.1.2. Population Characteristics

Flood events significantly influenced the population density and coverage of *B. planiculmis* ($p < 0.05$; Figure 4). Under SF conditions, the population density and coverage were recorded at 554.00 ± 169.14 individuals m^{-2} and 52.36% ± 21.32%, respectively. These values represent increases of 76.10% and 66.70%, respectively, compared with those observed under CF conditions. The average population heights were 36.83 ± 17.12 cm in SF conditions and 36.15 ± 18.31 cm in CF conditions, with no statistically significant difference ($p > 0.05$) detected between the two flooding conditions.

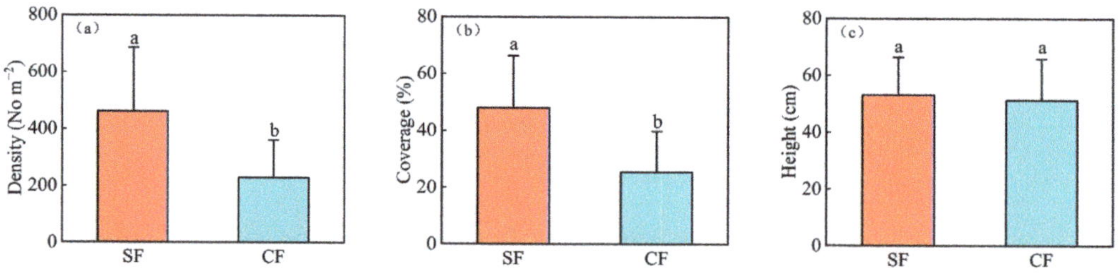

Figure 4. Ecological characteristics of *Bolboschoenus planiculmis* populations under different flooding conditions (mean ± standard error, n = 15): SF—seasonal flooding condition; CF—continuous flooding condition; (**a**) population density; (**b**) population coverage; and (**c**) population height. Different letters represent significant differences ($p < 0.05$).

3.1.3. Plant Traits Disparities

The findings from the biomass survey indicated significant differences in the aboveground, root, and bulb biomass per unit area of *B. planiculmis* under the different flooding conditions ($p < 0.05$; Figure 5). In CF conditions, the measured values for aboveground biomass, root biomass, and bulb biomass were 130.025 ± 10.50 g m^{-2}, 211.12 ± 12.20 g m^{-2}, and 114.32 ± 8.90 g m^{-2}, respectively. Conversely, under SF conditions, these values were significantly elevated: they were recorded at 345.37 ± 15.30 g m^{-2}, 315.80 ± 14.40 g m^{-2}, and 332.67 ± 16.70 g m^{-2}. The biomass of each component under SF conditions was 2.66, 1.50, and 2.91 times greater than that observed under CF conditions, respectively.

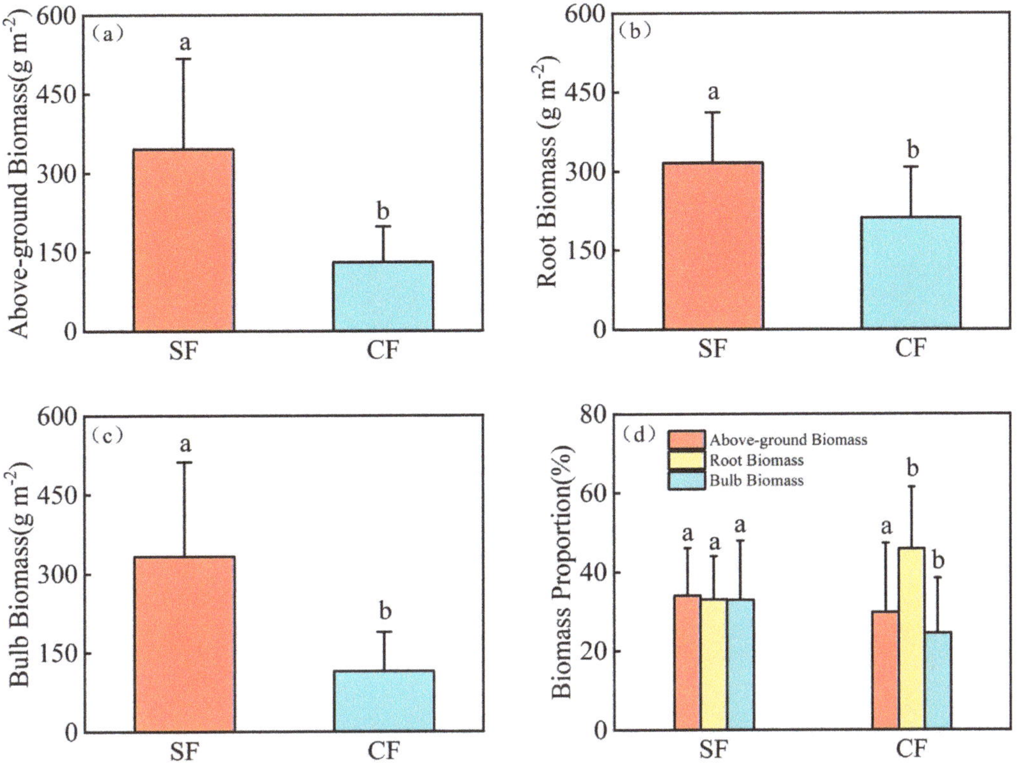

Figure 5. Characteristics of *Bolboschoenus planiculmis* biomass under different flooding conditions (mean ± standard error, n = 15). SF—seasonal flooding condition; CF—continuous flooding condition; (**a**) aboveground biomass (AB); (**b**) root biomass (RB); (**c**) bulb biomass (BB); (**d**) proportion of total biomass represented by each component. Different letters indicate significant differences in data under varying flood conditions ($p < 0.05$).

3.2. Analysis of Environmental Variables

3.2.1. Differences in Environmental Variables Under Different Flood Conditions

Environmental variables, including WL, pH, BD, SP, and MC, exhibited statistically significant differences across the hydrological conditions ($p < 0.05$; Table 1). Specifically, pH, WL, BD, SP, and MC were significantly higher under CF conditions than under SF conditions. The mean BD was recorded at 1.31 ± 0.11 g cm^{-3} under CF conditions, which exceeded the 1.06 ± 0.09 g cm^{-3} observed under SF conditions. Furthermore, the mean water content was 45.00 ± 3.00% under CF conditions, which was significantly higher than the 37 ± 2.7% observed under SF conditions. The BD and water content under CF conditions were 1.23 and 1.22 times greater than those observed under SF conditions, respectively. However, the EC and PD did not exhibit significant differences ($p > 0.05$) between the two flooding conditions.

Table 1. Soil physical properties of *Bolboschoenus planiculmis* community samples under different flooding conditions.

Flooding Conditions	SF	CF
WL (cm)	21.72 ± 1.68 [a]	32.72 ± 2.85 [b]
EC (mS cm^{-1})	731.70 ± 148.68 [a]	729.90 ± 143.25 [a]
pH	10.21 ± 0.20 [a]	10.41 ± 0.20 [b]
PD (g cm^{-3})	2.46 ± 0.41 [a]	2.57 ± 0.08 [a]
BD (g cm^{-3})	1.06 ± 0.91 [a]	1.31 ± 0.11 [b]
SP (%)	49.49 ± 6.40 [a]	61.81 ± 14.13 [b]
MC (%)	37.00 ± 2.70 [a]	45.00 ± 3.00 [b]

Notes: WL—water level; EC—electrical conductivity; PD—soil particle density; BD—soil bulk density; SP—soil porosity; MC—moisture content. Different letters indicate significant differences ($p < 0.05$).

3.2.2. Correlation Analysis

Spearman correlation analysis was conducted to examine the relationship between multi-level ecological characteristics and environmental variables (Figure 6). The results showed that BD and MC were significantly positively correlated with environmental factors ($p < 0.05$). The Shannon–Wiener and Simpson indices were significantly positively correlated with WL ($p < 0.05$). At the population level, a significant negative correlation was found between C and D with WL, PD, BD, SP, and MC ($p < 0.05$). At the plant trait level, AB, BB, and TB were significantly negatively correlated with WL, BD, and MC ($p < 0.05$). Based on these findings, WL, BD, and MC are key environmental factors influencing multi-level ecological characteristics.

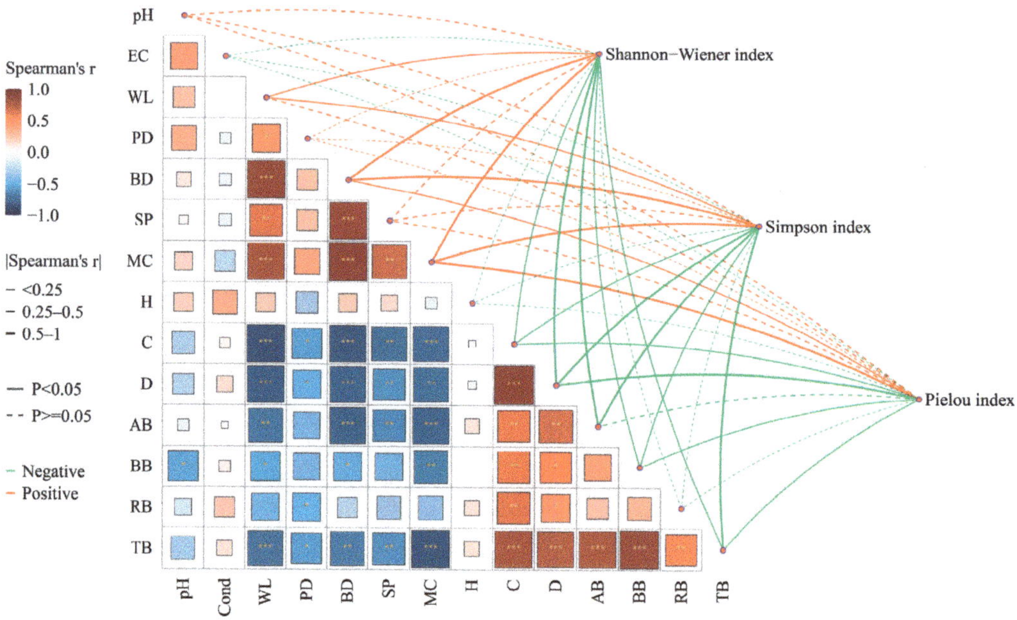

Figure 6. Correlation analysis of community species diversity with soil environmental indicators and community and plant ecological traits. WL—water level; EC—electrical conductivity; PD—soil particle density; BD—bulk density; SP—soil porosity; MC—moisture content; H—height; C—coverage; D—density; AB—aboveground biomass; BB—bulb biomass; RB—root biomass; TB—total biomass. * indicate significant differences at $p < 0.05$, ** indicate significant differences at $p < 0.01$, and *** indicate significant differences at $p < 0.001$.

3.3. Results of RDA

The results of the RDA (Figure 7) indicated that soil environmental characteristics accounted for 86.46% of the variation observed in the ecological characteristics of the *B. planiculmis* populations (Monte Carlo test, repeated 499 times; F = 11.0, p = 0.002). Both simple and conditional analyses indicated that WL is the most important influencing factor (Table 2). The results of the conditional effects analysis demonstrated that the influence of WL on the ecological characteristics of the *B. planiculmis* populations reached a significance level of 71% (F = 44.1, p = 0.002). These findings suggest that changes in water level, particularly those induced by flooding, have a significant impact on the ecological characteristics of *B. planiculmis* populations.

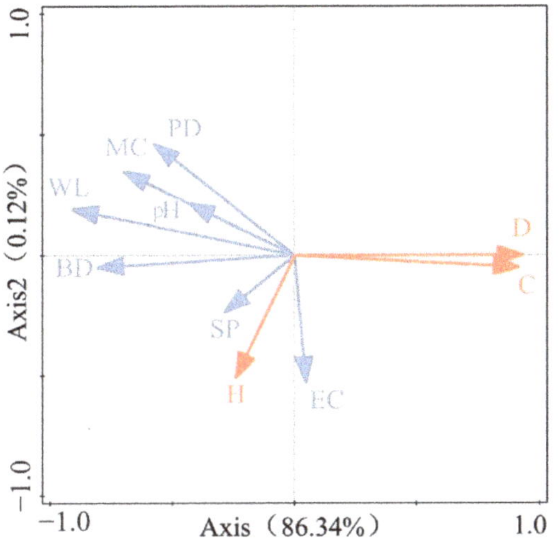

Figure 7. Results of RDA (conditional effects analysis). Blue arrows indicate population ecological characteristics and red arrows denote soil environmental indicators. WL—water level; EC—electrical conductivity; pH—pH level; PD—soil particle density; BD—bulk density; SP—soil porosity; MC—moisture content.

Table 2. RDA simple effects analysis with conditional effects analysis results.

Traits	Simple Effects Analysis			Conditional Effects Analysis		
	Explains %	Pseudo-F	p	Explains %	Pseudo-F	p
WL	65.7	34.5	0.002	71	44.1	0.002
BD	4.3	2.4	0.094	0.7	0.6	0.47
pH	1.7	0.9	0.452	4.8	4.4	0.064
EC	2.1	1.2	0.32	1.5	1	0.348
MC	2.3	1.4	0.274	2.8	1.9	0.164
SP	1.8	1	0.38	3.3	2.2	0.19
PD	1.8	1	0.414	2.4	1.8	0.174

Notes: WL—water level; EC—electrical conductivity; pH—pH level; PD—soil particle density; BD—bulk density; SP—soil porosity; MC—moisture content.

4. Discussion

4.1. Community Structure

Flooding events represent significant abiotic stresses that affect the development of wetland plant communities. Among these factors, the duration of flooding exerts a considerable influence on the structure of the plant community [44,51,52]. Our findings indicate that the species composition of *B. planiculmis* communities varied significantly across different flooding durations. Specifically, the number of species within the *B. planiculmis* community under CF conditions was three times greater than that observed under SF conditions.

Flooding rapidly transformed the habitat of *B. planiculmis* in SF environments, from wet to inundated conditions. According to the water gauge, the water level during the day of the flood ranged from 0 to 3.2 cm. This alteration in the hydrological environment has resulted in the loss of ecological niches for certain mesophytes and terrestrial plants, leading to changes in community species composition and a decline in species diversity [53]. In Dongting Lake, the onset of floods hindered the root respiration and nutrient exchange of *Carex*, leading to a decline in its dominance and a shift in the community's dominant species [54]. Notably, despite reduced biodiversity under SF conditions, *B. planiculmis* showed a higher degree of dominance within the community. This phenomenon is advantageous for the conservation of wetlands of *B. planiculmis*.

Our study identified a significant positive correlation ($p < 0.05$) between the water level and Shannon's diversity index [55,56]. Fluctuations in water levels directly influence the restructuring of root tissues, growth of stems and leaves, and accumulation of photosynthetic pigments, which have a comparatively smaller impact on plant communities under CF conditions [57,58]. In CF areas, plants are subjected to consistent flooding, which promotes the development of well-structured root aeration tissues that facilitate their adaptation to flooding through various physiological adjustments. In contrast, in the SF areas, the land remains moist without inundation until flooding occurs. The abrupt onset of flooding results in a rapid increase in water levels, placing plants under more challenging conditions than those in CF areas. This phenomenon, referred to as rough flood abrupt alteration (DFAA), may disrupt the balance of endogenous growth hormones in plants [59]. Increased ACC synthase activity promotes the conversion of S-adenosylmethionine (SAM) to 1-aminocyclopropane-1-carboxylate (ACC). Simultaneously, the activity of ACC oxidase increases, resulting in the enhanced oxidation of ACC to ethylene [60,61]. The consequent increase in the production of endogenous ethylene, along with abnormal concentrations, amplifies ethylene signaling, which can inhibit root system development and exacerbate plant hypoxia during the abrupt transition from aerobic to anaerobic conditions [62]. This hypoxic condition is further intensified by the sudden shift from aerobic to anaerobic environments [63,64]. At the same time, studies have shown that the biosynthesis of brassinosteroids, auxin, and gibberellins (GAs) is generally down-regulated in flooded plants. This leads to the inhibition of root growth and lateral expansion, which can adversely affect plant development and overall growth in waterlogged conditions [65]. Flood events significantly influenced the species composition and physiological responses of the *B. planiculmis* community, with prolonged flooding enhancing community diversity and promoting plant adaptation.

4.2. Population Ecological Characteristics

The results of this study revealed a significant negative correlation between water level and both population density and coverage of *B. planiculmis*, consistent with the findings of previous studies [66,67]. An increase in the water level is recognized as one of the most direct indicators of flooding, as the influx of floodwater alters soil structure and oxygen availability, which can impede the growth and spread of *B. planiculmis* [68]. Furthermore, flood inundation suppresses the photosynthetic capacity of *B. planiculmis*, resulting in decreased biomass accumulation and reproductive potential [30]. Additionally, flood disturbances exacerbate competitive pressure between *B. planiculmis* and other flood-tolerant plant species [69]. The interplay between these factors significantly reduced the population density and coverage of *B. planiculmis*.

However, it is noteworthy that water level did not exert a significant influence on the height of the *B. planiculmis* community, which may be attributed to the species' tolerance to varying water levels. Ding et al. indicated that mature individual *B. planiculmis* can thrive in water depths of up to 45 cm, which was why the height of the community remained largely unaffected by low-water flooding conditions [70]. In contrast, Yang et al. observed that flood depths exceeding 20 cm had a pronounced impact on *B. planiculmis* seedlings, with water levels recorded at 21.72 ± 1.68 cm in SF conditions and 32.72 ± 2.85 cm in

CF conditions [71]. This suggests that elevated water levels under CF conditions impose additional stress on seedling growth, leading to a significant reduction in the density and coverage of *B. planiculmis* communities. Consequently, the heightened water levels in CF areas exerted greater stress on the growth of *B. planiculmis* seedlings, resulting in a markedly lower density and coverage than those observed under seasonal flooding conditions.

4.3. Plant Traits

At the vegetative level, flooding significantly influences plant growth, prompting plants to adopt either stationary or escape strategies in response to inundation [72,73]. Plants use various strategies to adapt to changes in hydrological conditions, including root restructuring [74] and upward branching to access oxygen [75]. Biomass allocation has been evaluated as an indicator of plant traits because the manner in which plants allocate biomass in response to environmental stress directly reflects their survival mechanisms [76,77].

In our comparative analysis of the ratios of aboveground, bulb, and root biomass to total biomass, we observed that the difference in the proportion of aboveground biomass between the two flooding conditions was not statistically significant. However, the SF condition resulted in a higher proportion of bulb biomass, whereas the CF condition resulted in a higher root biomass ratio. This discrepancy is attributed to the prolonged exposure of plants to flooding under CF conditions, which enhances the allocation of biomass to the root system [27]. Furthermore, when examining total biomass and other biomass metrics, the productivity per unit area of *B. planiculmis* communities under SF conditions was greater than under CF conditions. This finding suggests that a seasonal inundation strategy is more effective than a continuous inundation strategy for restoring habitats that serve as food sources for rare avian species. Consistent with our findings, research conducted on tussock grass wetlands in the Momoge region has identified seasonal flooding as a critical hydrological regulation method for wetland restoration [78].

4.4. Hydrological Regulation Strategies for B. planiculmis Wetlands

In the context of global climate change and the escalating effects of anthropogenic activities, the restoration and conservation of wetlands in frequently inundated floodplain regions are of paramount importance. Ecological engineering strategies constitute a vital approach to restoration because the regulation of wetland hydrology plays a crucial role in these initiatives [79]. Our research indicates that although plant communities demonstrate greater diversity under CF conditions, the population characteristics and biomass allocation strategies of *B. planiculmis* are more advantageous under SF conditions.

Building on previous MNNR studies related to hydrological regulation projects, wetland restoration is crucial for protecting the *Grus leucogeranus* stopover [32]. We propose that seasonal inundation will enhance the conservation and management of wetlands that support *B. planiculmis*. Accordingly, we recommend the following measures: Hydrological regulation in CF areas, which will involve enhancing hydrological connectivity through engineering interventions to transform CF areas into more conducive environments for *B. planiculmis*. This strategy aims to increase the density and biomass of *B. planiculmis*, particularly its bulb productivity, which is a critical food source for rare bird species. Flood-level reduction: this will involve implementing measures to mitigate excessively high flood levels that can cause significant damage to *B. planiculmis* wetlands. A judicious reduction in water levels will not only facilitate the protection of *B. planiculmis* but will also contribute to the overall conservation of the wetland ecosystem. Mitigation of drought–flood impacts: Engineering solutions will be utilized to alleviate the rapid effects of flooding, thereby creating more favorable conditions for plants to adapt to environmental changes and adjust their growth strategies. This approach also supports the preservation of species diversity within communities.

The goal of hydrological control measures is to achieve near-natural wetland restoration, which is a challenging objective. On one hand, restoration requires long-term regulation; on the other hand, it is crucial to precisely determine the scale of hydrological adjustments to ensure the accuracy of the restoration process.

5. Conclusions

This study investigated the ecological characteristics of wetlands of *B. planiculmis* at various levels under different flooding conditions. The results indicated that flooding events significantly influenced the structure and diversity of *B. planiculmis* communities, as well as their ecological characteristics and plant traits. Correlation and RDA analyses revealed that variations in water levels due to flooding conditions were the primary environmental factors driving the differences in the ecological characteristics of the *B. planiculmis* wetlands at multiple scales. In conclusion, during the near-natural restoration of *B. planiculmis* wetlands, the implementation of hydrological control measures such as enhancing the hydrological environment, reducing flood levels, and alleviating the effects of rapid flooding can facilitate the creation of a more favorable environment for *B. planiculmis* and contribute to the protection of *B. planiculmis* wetlands and habitats of rare bird species. The study provides valuable insights and restoration strategies for wetlands worldwide that have suffered habitat degradation due to flood events. It also emphasizes the importance of adopting targeted, sustainable approaches to hydrological regulation in order to restore ecosystem functions and strengthen long-term environmental resilience, ultimately achieving near-natural restoration. Currently, there are still gaps in hydrological control strategies, and more detailed approaches are needed based on actual field conditions. Future efforts should focus on long-term observations to better understand the impact of flood events on the B. planiculmis community. Additionally, strengthening collaboration with wetland protection organizations is essential for achieving the successful restoration of wetland ecological functions.

Author Contributions: Conceptualization, M.Z. and L.C.; methodology, L.C., M.Z. and J.Z.; software, L.C. and Y.X.; validation, L.C., M.Z. and J.Z.; formal analysis, M.Z.; investigation, L.C.; resources, M.Z.; data curation, L.C.; writing—original draft preparation, L.C., M.Z. and Y.X.; writing—review and editing S.T., Y.A., M.Z. and C.Z.; visualization, L.C.; supervision, M.Z.; project administration, M.Z.; funding acquisition, M.Z. All authors have read and agreed to the published version of the manuscript.

Funding: This research was financially supported by the National Key Research and Development Program of China (No. 2023YFF1304502), the National Natural Science Foundation of China (No. 42401132), the China Postdoctoral Science Foundation (No. 2024M753210), the Postdoctoral Fellowship Program of CPSF (No. GZC20232639), the Major Scientific and Technological Project of Jilin Province of China (No. 20230303005SF), the Science and Technology Youth talent Support Project of Jilin Province (No. QT202411), and the Science and Technology Development Planning Project of Jilin Province (No. YDZJ202401477ZYTS).

Data Availability Statement: The data adopted in the study are included in the article; further inquiries can be directed to the corresponding author.

Conflicts of Interest: The authors declare no conflicts of interest.

References

1. Jongman, B. Effective adaptation to rising flood risk. *Nat. Commun.* **2018**, *9*, 1986. [CrossRef] [PubMed]
2. Prigent, C.; Matthews, E.; Aires, F.; Rossow, W.B. Remote sensing of global wetland dynamics with multiple satellite data sets. *Geophys. Res. Lett.* **2001**, *28*, 4631–4634. [CrossRef]
3. Tellman, B.; Sullivan, J.A.; Kuhn, C.; Kettner, A.J.; Doyle, C.S.; Brakenridge, G.R.; Slayback, D.A. Satellite imaging reveals increased proportion of population exposed to floods. *Nature* **2021**, *596*, 80–86. [CrossRef]
4. Brody, S.D.; Peacock, W.G.; Gunn, J. Ecological indicators of flood risk along the Gulf of Mexico. *Ecol. Indic.* **2012**, *18*, 493–500. [CrossRef]
5. Liu, J.; Engel, B.A.; Zhang, G.; Wang, Y.; Wu, Y.; Zhang, M.; Zhang, Z. Hydrological connectivity: One of the driving factors of plant communities in the Yellow River Delta. *Ecol. Indic.* **2020**, *112*, 106150. [CrossRef]
6. Marazzi, L.; Gaiser, E.E.; Jones, V.J.; Tobias, F.A.; Mackay, A.W. Algal richness and life-history strategies are influenced by hydrology and phosphorus in two major subtropical wetlands. *Freshwater Biol.* **2017**, *62*, 274–290. [CrossRef]

7. Meng, Y.; Yu, S.; Yu, Y.; Jiang, L. Flooding depth and duration concomitantly influence the growth traits and yield of rice. *Irrig. Drain.* **2022**, *71*, 94–107. [CrossRef]
8. Åhlén, I.; Thorslund, J.; Hambäck, P.; Destouni, G.; Jarsjö, J. Wetland position in the landscape: Impact on water storage and flood buffering. *Ecohydrology* **2022**, *15*, e2458. [CrossRef]
9. Zedler, J.B.; Kercher, S. Wetland resources: Status, trends, ecosystem services, and restorability. *Annu. Rev. Environ. Resour.* **2005**, *30*, 39–74. [CrossRef]
10. Wu, Y.; Zhang, G.; Rousseau, A.N.; Xu, Y.J.; Foulon, É. On how wetlands can provide flood resilience in a large river basin: A case study in Nenjiang River Basin, China. *J. Hydrol.* **2020**, *587*, 125012. [CrossRef]
11. Deschamps, L.; Proulx, R.; Rheault, G.; Gross, N.; Watson, C.; Maire, V. Species richness drives selection of individuals within wetlands based on traits related to acquisition and utilization of light. *Ecol. Evol.* **2023**, *13*, e9959. [CrossRef] [PubMed]
12. Nilsson, C.; Brown, R.L.; Jansson, R.; Merritt, D.M. The role of hydrochory in structuring riparian and wetland vegetation. *Biol. Rev.* **2010**, *85*, 837–858. [CrossRef]
13. Allen, B.E.; Azeria, E.T.; Bried, J.T. Linking functional diversity, trait composition, invasion, and environmental drivers in boreal wetland plant assemblages. *J. Veg. Sci.* **2021**, *32*, e13073. [CrossRef]
14. Li, W.; Tan, R.; Yang, Y.M.; Wang, J. Plant diversity as a good indicator of vegetation stability in a typical plateau wetland. *J. Mountain Sci.* **2014**, *11*, 464–474. [CrossRef]
15. Dronova, I.; Taddeo, S.; Harris, K. Plant diversity reduces satellite-observed phenological variability in wetlands at a national scale. *Sci. Adv.* **2022**, *8*, eabl8214. [CrossRef] [PubMed]
16. Engelhardt, K.A.M.; Ritchie, M.E. Effects of macrophyte species richness on wetland ecosystem functioning and services. *Nature* **2001**, *411*, 687–689. [CrossRef] [PubMed]
17. Fons, V.D.P. Biodiversity and ecosystem functioning in naturally assembled communities. *Biol. Rev.* **2019**, *94*, 1220–1245.
18. Kajtoch, Ł.; Grzędzicka, E.; Piechnik, Ł.; Wyka, J.; Lešo, P. Evergreen ivy vines as a key element maintaining the high diversity of birds wintering in Central European forests. *For. Ecol. Manag.* **2023**, *544*, 121165. [CrossRef]
19. Souza, L.; Zelikova, T.J.; Sanders, N.J. Bottom–up and top–down effects on plant communities: Nutrients limit productivity, but insects determine diversity and composition. *Oikos* **2016**, *125*, 566–575. [CrossRef]
20. Woodcock, B.A.; Bullock, J.M.; Nowakowski, M.; Orr, R.; Tallowin, J.R.B.; Pywell, R.F. Enhancing floral diversity to increase the robustness of grassland beetle assemblages to environmental change. *Conserv. Lett.* **2012**, *5*, 459–469. [CrossRef]
21. Geppert, C.; Perazza, G.; Wilson, R.J.; Bertolli, A.; Prosser, F.; Melchiori, G.; Marini, L. Consistent population declines but idiosyncratic range shifts in Alpine orchids under global change. *Nature Commun.* **2020**, *11*, 5835. [CrossRef]
22. Yang, C.; Shen, X.; Wu, J.; Shi, X.; Cui, Z.; Tao, Y.; Huang, Q. Driving forces and recovery potential of the macrophyte decline in East Taihu Lake. *J. Environ. Manag.* **2023**, *342*, 118154. [CrossRef] [PubMed]
23. Yuan, S.; Yang, Z.; Liu, X.; Wang, H. Key parameters of water level fluctuations determining the distribution of Carex in shallow lakes. *Wetlands* **2017**, *37*, 1005–1014. [CrossRef]
24. Zhang, M.; Liu, S.; Zhang, D.; Qi, Q.; An, Y.; Cui, G.; Tong, S. Effects of hydrological regimes on soil seed banks in the Carex hummock in wetlands: Implications for restoration. *Ecol. Eng.* **2024**, *202*, 107219. [CrossRef]
25. Ayi, Q.; Zeng, B.; Liu, J.; Li, S.; van Bodegom, P.M.; Cornelissen, J.H. Oxygen absorption by adventitious roots promotes the survival of completely submerged terrestrial plants. *Ann. Bot.* **2016**, *118*, 675–683. [CrossRef] [PubMed]
26. Luo, F.L.; Jiang, X.X.; Li, H.L.; Yu, F.H. Does hydrological fluctuation alter impacts of species richness on biomass in wetland plant communities? *J. Plant Ecol.* **2016**, *9*, 434–441. [CrossRef]
27. Zhang, Q.; Visser, E.J.; de Kroon, H.; Huber, H. Life cycle stage and water depth affect flooding-induced adventitious root formation in the terrestrial species Solanum dulcamara. *Ann. Bot.* **2015**, *116*, 279–290. [CrossRef] [PubMed]
28. An, Y.; Song, T.; Zhang, Y.; Tong, S.; Liu, B. Optimum water depth for restoration of *Bolboschoenus planiculmis* in wetlands in semi-arid regions. *Hydrobiologia* **2022**, *849*, 13–28. [CrossRef]
29. Yu, D.; Tang, H.; Li, P.; Zhou, M.; Zhao, G.; Lou, Y. Temperature and flooding depth thresholds for early recruitment stages in a bulbous plant *Bolboschoenus planiculmiss*. *Wetlands Ecol. Manag.* **2023**, *31*, 19–30. [CrossRef]
30. An, Y.; Gao, Y.; Tong, S. Emergence and growth performance of *Bolboschoenus planiculmis* varied in response to water level and soil planting depth: Implications for wetland restoration using tuber transplantation. *Aquat. Bot.* **2018**, *148*, 10–14. [CrossRef]
31. Kim, G.Y.; Kim, J.Y.; Ganf, G.G.; Lee, C.W.; Joo, G.J. Impact of over-wintering waterfowl on tuberous bulrush (*Bolboschoenus planiculmis*) in tidal flats. *Aquat. Bot.* **2013**, *107*, 17–22. [CrossRef]
32. Jiang, H.; Wen, Y.; Zou, L.; Wang, Z.; He, C.; Zou, C. The effects of a wetland restoration project on the Siberian crane (*Grus leucogeranus*) population and stopover habitat in Momoge National Nature Reserve, China. *Ecol. Eng.* **2016**, *96*, 170–177. [CrossRef]
33. Ding, S.; Yu, X.; Zhang, J.; Yin, Z.; Zou, Y.; Wang, G.; He, C. Bioconcentration and translocation of elements regulate plant responses to water-salt conditions in saline-alkaline wetlands. *Environ. Exp. Bot.* **2021**, *183*, 104360. [CrossRef]
34. Zhang, L.; Zhang, G.; Li, H.; Sun, G. Eco-physiological responses of *Scirpus planiculmis* to different water-salt conditions in Momoge Wetland. *Polish J. Environ. Stud.* **2014**, *23*, 1813–1820.
35. Qi, Q.; Zhang, D.; Zhang, M.; Tong, S.; Wang, W.; An, Y. Spatial distribution of soil organic carbon and total nitrogen in disturbed *Carex tussock* wetland. *Ecol. Indic.* **2021**, *120*, 106930. [CrossRef]

36. Wang, Y.; Gong, M.; Zou, C.; Zhou, T.; Wen, W.; Liu, G.; Tao, W. Habitat selection by Siberian Cranes at their core stopover area during migration in Northeast China. *Global Ecol. Conserv.* **2022**, *33*, e01993. [CrossRef]
37. Zhao, Y.; Wang, G.; Zhao, M.; Wang, M.; Jiang, M. Direct and indirect effects of soil salinization on soil seed banks in salinizing wetlands in the Songnen Plain, China. *Sci. Total Environ.* **2022**, *819*, 152035. [CrossRef] [PubMed]
38. Zhang, M.; Zhang, D.; Qi, Q.; Tong, S.; Wang, X.; An, Y.; Lu, X. Flooding effects on population and growth characteristics of *Bolboschoenus planiculmis* in Momoge wetland, northeast China. *Ecol. Indic.* **2022**, *137*, 108730. [CrossRef]
39. An, Y.; Gao, Y.; Zhang, Y.; Tong, S.; Liu, X. Early establishment of Suaeda salsa population as affected by soil moisture and salinity: Implications for pioneer species introduction in saline-sodic wetlands in Songnen Plain, China. *Ecol. Indic.* **2019**, *107*, 105654. [CrossRef]
40. Xin, Y.; Qi, Q.; Zhang, M.; Zhang, D.; Cui, G.; An, Y.; Xiang, X. Responses of plant productivity and diversity to drought in *Carex schmidtii* tussock wetlands, Northeast China. *Community Ecol.* **2024**, *25*, 1–13. [CrossRef]
41. Zhang, D.; Qi, Q.; Wang, X.; Tong, S.; Lv, X.; An, Y.; Zhu, X. Physiological responses of *Carex schmidtii* Meinsh to alternating flooding-drought conditions in the Momoge wetland, northeast China. *Aquat. Bot.* **2019**, *153*, 33–39. [CrossRef]
42. Wu, Z. *Vegetation of China*; Science Press: Beijing, China, 1980.
43. Zhang, M.; Qi, Q.; Zhang, D.; Tong, S.; Wang, X.; An, Y.; Lu, X. Effect of priming on *Carex schmidtii* seed germination and seedling growth: Implications for tussock wetland restoration. *Ecol. Eng.* **2021**, *171*, 106389. [CrossRef]
44. Qi, Q.; Zhang, D.; Zhang, M.; Tong, S.; An, Y.; Wang, X.; Zhu, G. Hydrological and microtopographic effects on community ecological characteristics of Carex schmidtii tussock wetland. *Sci. Total Environ.* **2021**, *780*, 146630. [CrossRef]
45. Yu, Q.; Suo, L.; Qi, J.; Wang, Y.; Hu, Q.; Shan, Y.; Zhao, Y. Soil habitat condition shapes Tamarix chinensis community diversity in the coastal saline-alkali soils. *Front. Plant Sci.* **2023**, *14*, 1156297. [CrossRef] [PubMed]
46. IBM Corporation. *SPSS Statistics (Version 23) [Computer Software]*; IBM Corporation: Armonk, NY, USA, 2015.
47. R Core Team. *R: A Language and Environment for Statistical Computing (4.3.3)*; R Foundation for Statistical Computing: Vienna, Austria, 2024. Available online: https://www.r-project.org/ (accessed on 14 September 2024).
48. Microcomputer Power. *Canoco 5.0 [Computer Software]*; Microcomputer Power: Ithaca, NY, USA, 2012.
49. Microsoft Corporation. *Microsoft Excel (Version 2202) [Computer Software]*; Microsoft Corporation: Redmond, WA, USA, 2021. Available online: https://www.microsoft.com (accessed on 1 June 2022).
50. OriginLab Corporation. *Origin 2022 [Computer Software]*; OriginLab Corporation: Northampton, MA, USA, 2022. Available online: https://www.originlab.com (accessed on 1 September 2022).
51. Deane, D.C.; Nicol, J.M.; Gehrig, S.L.; Harding, C.; Aldridge, K.T.; Goodman, A.M.; Brookes, J.D. Hydrological-niche models predict water plant functional group distributions in diverse wetland types. *Ecol. Appl.* **2017**, *27*, 1351–1364. [CrossRef] [PubMed]
52. Moor, H.; Rydin, H.; Hylander, K.; Nilsson, M.B.; Lindborg, R.; Norberg, J. Towards a trait-based ecology of wetland vegetation. *J. Ecol.* **2017**, *105*, 1623–1635. [CrossRef]
53. Capon, S.J. Flood variability and spatial variation in plant community composition and structure on a large arid floodplain. *J. Arid Environ.* **2005**, *60*, 283–302. [CrossRef]
54. Huang, Y.; Chen, X.S.; Li, F.; Hou, Z.Y.; Li, X.; Zeng, J.; E, Y.H. Community trait responses of three dominant macrophytes to variations in flooding during 2011–2019 in a Yangtze River-connected floodplain wetland (Dongting Lake, China). *Front. Plant Sci.* **2021**, *12*, 604677. [CrossRef] [PubMed]
55. Valle Ferreira, L. Effects of flooding duration on species richness, floristic composition and forest structure in river margin habitat in Amazonian blackwater floodplain forests: Implications for future design of protected areas. *Biodivers. Conserv.* **2000**, *9*, 1–14. [CrossRef]
56. Yi, X.; Huang, Y.; Jiang, Y.; Ma, M.; Chen, Q.; Wu, S. Flooding Depth and Flooding Duration with the Zonation of Riparian Plant Communities in the Three Gorges Reservoir of China. *Water* **2023**, *15*, 3228. [CrossRef]
57. Spence, D.H.N. Factors controlling the distribution of freshwater macrophytes with particular reference to the Loch of Scotland. *J. Ecol.* **1967**, *55*, 147–170. [CrossRef]
58. Casanova, M.; Brock, M.A. How do depth, duration and frequency of flooding influence the establishment of wetland plant communities? Aust. *J. Ecol.* **2000**, *25*, 298–309.
59. Xiong, Q.; Deng, Y.; Zhong, L.; He, H.; Chen, X. Effects of drought-flood abrupt alternation on yield and physiological characteristics of rice. *Int. J. Agric. Biol.* **2018**, *20*, 1107–1116.
60. Adams, D.O.; Yang, S. Ethylene biosynthesis: Identification of 1-aminocyclopropane-1-carboxylic acid as an intermediate in the conversion of methionine to ethylene. *Proc. Natl. Acad. Sci. USA* **1979**, *76*, 170–174. [CrossRef]
61. Blom, C.W.P.M.; Voesenek, L.A.C.J. Flooding: The survival strategies of plants. *Trends Ecol. Evol.* **1996**, *11*, 290–295. [CrossRef] [PubMed]
62. Maric, A.; Hartman, S. Ethylene controls translational gatekeeping to overcome flooding stress in plants. *EMBO J.* **2022**, *41*, e112282. [CrossRef] [PubMed]
63. Visser, E.J.W.; Nabben, R.H.M.; Blom, C.W.P.M.; Voesenek, L.A.C.J. Elongation by primary lateral roots and adventitious roots during conditions of hypoxia and high ethylene concentrations. *Plant Cell Environ.* **1997**, *20*, 647–653. [CrossRef]
64. Fukaki, H.; Tasaka, M. Hormone interactions during lateral root formation. *Plant Mol. Biol.* **2009**, *69*, 437–449. [CrossRef] [PubMed]

65. Ruperti, B.; Botton, A.; Populin, F.; Eccher, G.; Brilli, M.; Quaggiotti, S.; Meggio, F. Flooding responses on grapevine: A physiological, transcriptional, and metabolic perspective. *Front. Plant Sci.* **2019**, *10*, 339. [CrossRef] [PubMed]
66. Zhu, G.; Li, Y.; Sun, Z.; Kanae, S. Response of vegetation to submergence along the Jiang Reach of the Yangtze River. *PLoS ONE* **2021**, *16*, e0251015.
67. Wang, Y.; Li, J.; Qian, K.; Ye, M. Response of plant species diversity to flood irrigation in the Tarim River Basin, Northwest China. *Sustainability* **2023**, *15*, 1243. [CrossRef]
68. Vervuren, P.J.A.; Blom, C.W.P.M.; De Kroon, H. Extreme flooding events on the Rhine and the survival and distribution of riparian plant species. *J. Ecol.* **2003**, *91*, 135–146. [CrossRef]
69. Liu, Y.; Ding, Z.; Bachofen, C.; Lou, Y.; Jiang, M.; Tang, X.; Buchmann, N. The effect of saline-alkaline and water stresses on water use efficiency and standing biomass of *Phragmites australis* and *Bolboschoenus planiculmis*. *Sci. Total Environ.* **2018**, *644*, 207–216. [CrossRef]
70. Ding, Z.; Liu, Y.; Lou, Y.; Jiang, M.; Li, H.; Lü, X. How soil ion stress and type influence the flooding adaptive strategies of *Phragmites australis* and *Bolboschoenus planiculmis* in temperate saline–alkaline wetlands. *Sci. Total Environ.* **2021**, *771*, 144654. [CrossRef] [PubMed]
71. Yang, H.; Kim, J.H.; Lee, E.J. Seasonal flooding regime effects on the survival, growth, and reproduction of *Bolboschoenus planiculmis* under East Asian monsoon. *Flora* **2021**, *285*, 151960. [CrossRef]
72. Striker, G.G.; Izaguirre, R.F.; Manzur, M.E.; Grimoldi, A.A. Different strategies of *Lotus japonicus*, *L. corniculatus*, and *L. tenuis* to deal with complete submergence at seedling stage. *Plant Biol.* **2012**, *14*, 50–55. [CrossRef] [PubMed]
73. Lan, Z.; Huang, H.; Chen, Y.; Liu, J.; Chen, J.; Li, L.; Chen, J. Testing mechanisms underlying responses of plant functional traits to flooding duration gradient in a lakeshore meadow. *J. Freshwater Ecol.* **2019**, *34*, 481–495. [CrossRef]
74. Potocka, I.; Szymanowska-Pułka, J. Morphological responses of plant roots to mechanical stress. *Ann. Bot.* **2018**, *122*, 711–723. [CrossRef]
75. Oesterheld, M.; McNaughton, S.J. Interactive effect of flooding and grazing on the growth of Serengeti grasses. *Oecologia* **1991**, *88*, 153–156. [CrossRef] [PubMed]
76. Hough-Snee, N.; Nackley, L.L.; Kim, S.H.; Ewing, K. Does plant performance under stress explain divergent life history strategies? The effects of flooding and nutrient stress on two wetland sedges. *Aquatic Bot.* **2015**, *120*, 151–159. [CrossRef]
77. Roa-Fuentes, L.L.; Campo, J.; Parra-Tabla, V. Plant biomass allocation across a precipitation gradient: An approach to seasonally dry tropical forest at Yucatán, Mexico. *Ecosystems* **2012**, *15*, 1234–1244. [CrossRef]
78. Zhang, D.; Qi, Q.; Tong, S.; Wang, X.; An, Y.; Zhang, M.; Lu, X. Soil degradation effects on plant diversity and nutrient in tussock meadow wetlands. *J. Soil Sci. Plant Nutr.* **2019**, *19*, 535–544. [CrossRef]
79. Dunlop, T.; Glamore, W.; Felder, S. Restoring estuarine ecosystems using nature-based solutions: Towards an integrated eco-engineering design guideline. *Sci. Total Environ.* **2023**, *873*, 162362. [CrossRef] [PubMed]

water

MDPI

Article

Spatial Differences in Soil Nutrients Along a Hydrographic Gradient on Floodplains in Dongting Lake

Jiayi Li [1], Yuanmi Wu [1], Dong Peng [2], Mingzhu Chen [1,3], Lingli Peng [1,4], Beth A. Middleton [5] and Ting Lei [1,6,*]

1 School of Ecology and Nature Conservation, Beijing Forestry University, Beijing 100083, China; lijy2000@bjfu.edu.cn (J.L.)
2 School of Geography and Ocean Science, Nanjing University, Nanjing 210023, China
3 Lin Zhiyuan (Beijing) Forestry Engineering Consulting Co., Ltd., Beijing 100013, China
4 Academy of Forest and Grassland Inventory and Planning, National Forestry and Grassland Administration, Beijing 100013, China
5 Wetland and Aquatic Research Center, U.S. Geological Survey, Lafayette, LA 70506, USA; middletonb@usgs.gov
6 National Field Scientific Observation and Research Station of Dongting Lake Wetland Ecosystem, Changde 415904, China
* Correspondence: leiting@bjfu.edu.cn

Abstract: The spatial heterogeneity of soil nutrients is crucial for the water bird and whole floodplain wetland ecosystem in large lakes, and it is influenced by the dramatic water level changes and sedimentation progress in West Dongting Lake (WDL). Soil samples were collected at various soil depths along the Yuan River and Li River that feed into WDL. The concentrations of soil total organic carbon (TOC), total nitrogen (TN), total phosphorus (TP), and soil grain size were tested. The stoichiometric ratios of C, N, P, and the mean value of soil grain size (*Mz*) were calculated. The differences of soil TOC, TN, TP and the stoichiometric ratio at different sites and soil depths were compared. Linear regression was used to explore the relationships of *Mz* and nutrient concentrations, and relationships between TOC, TN, and TP. Redundancy analysis was used to explore the relationship between soil nutrients, heavy metal concentrations, and plant community diversity. The results showed that the distributions of soil TOC, TN, and TP concentrations differed across regions in west Dongting Lake along the Yuan and Li Rivers. Total organic carbon concentration differed at different sedimentation depths. Soil grain size showed negative effect with soil TOC, TN, and TP concentrations in this region. Plant community diversity correlated positively with soil TOC and negatively with Hg. West Dongting Lake was N limited despite the high wet deposition of N. It could potentially be attributed to the insufficient presence of aerobic environments for microbes during intermittent flooding of the floodplain, coupled with feeble mineralization. This study can provide valuable insights for the conservation of water bird habitats and wetland ecosystems.

Keywords: floodplain; soil nutrients; spatial difference; soil grain size; west Dongting Lake

1. Introduction

The West Dongting Lake (WDL) is a part of Dongting Lake, which is the second largest freshwater lake in China. It is also an important water bird habitat on the East Asian-Australasian Flyway and was designated as a wetland of international importance by the Ramsar Convention in 2002 [1]. The nutrients of WDL play a critical role for supporting wetland ecosystem fitness and water bird habitat health. As a part of Dongting Lake, the second largest freshwater lake in China that is directly connected to the Yangtze River, West Dongting Lake is a floodplain lake with strong annual water level fluctuations and functions as a floodwater buffer.

West Dongting Lake is situated at a higher elevation with only one outlet leading to the South Dongting Lake. However, it receives inflows from the Yuan River and Li

River, resulting in a high rate of sediment deposition [2]. Regional nutrient cycles regulate the feedback of trophic ecosystems [3,4], which may be influenced by global and regional changes in temperature and/or moisture environments [5–7]. Soil total organic carbon (TOC) has a significant effect on regulating the potential denitrification rate [8], and long-term N removal could influence the production of wetland soil organic carbon [9]. The stoichiometric characteristics of soil total organic carbon (TOC), total nitrogen (TN), and total phosphorus (TP) can be greatly influenced by sedimentation level, water dynamics, sediment availability, and suspended particles [10]. After the completion and onset of the full operation of the Three Gorges Dam in China, the hydrological and sedimentation patterns changed in WDL [2,11–13]. Thus, nutrient distribution patterns on the monsoonal floodplain of WDL are affected by the Three Gorges Dam and may not be similar to the patterns of temperate floodplains.

Floodplain is defined as the maximum extent of flood in this study, receiving floods annually or biannually [14]. The concentration of soil TOC depends on soil properties such as depth, soil moisture, and salinity levels during flooding following monsoonal rains in these floodplains [15]. Soil particle size greatly impacts nutrient concentration and availability via losses in organic carbon and nutrients [16,17]. Grain size can reflect sedimentation processes, i.e., the movement of sediment before deposition as well as the intrinsic properties of the sediment itself [18–20]. Also, nutrient levels in soil can be influenced by the eutrophication levels of tributaries-feeding wetlands [21]. Seasonal flood flow from upstream and its interaction with floodplain geomorphology may bring a sudden flux of nutrients between a river channel and its floodplain [22], particularly in locations where water flow is slow and sediment deposition is high [23,24].

West Dongting Lake could act as a storage pool for nutrients and sediment in the Yangtze River watershed. It has historically had large amounts of sedimentation [25] carried by the main tributaries (Yuan River and Li River). While not well understood in these monsoonal floodplains, hydrological patterns can shift the form, transformation, and distribution patterns of nutrients in wetlands [15,22,26–31]. Previous studies have not adequately examined the distribution of soil nutrients across landscapes, especially in monsoonal floodplains like those near Dongting Lake.

Nutrient stoichiometry can reflect the health of plant communities, the soil environment, and even the trends of ecosystem change in wetlands [32]. The soil carbon–nitrogen ratio is an important indicator of the soil organic matter decomposition [33,34]. Stoichiometric patterns of carbon, nitrogen, and phosphorus were studied previously in wetland soils in the Sanjiang Plain of northeast China [32] and estuarine soils of Shuangtaizi Estuary in northeast China [35]. However, the relationship between nutrients and hydrology in floodplain wetland of large lake may be more complicated in WDL. Research on the spatial distribution of soil nutrients along the tributaries waterflow of Dongting Lake is still insufficient, except for a study on small-scale patterns in East Dongting Lake [36].

The sampling sites near the Yuan River and Li River flood channels can be affected by flow-sediment regulation related to the operation of the upstream Three Gorges Dam of the Yangtze River [2]. The velocity of water flowing from the two tributaries decreases after flowing into the inlet of the lake, resulting in the deposition of sediment near the delta. Land reclamation and siltation have reduced the size of the lake [37]. In the last half-decade, Dongting Lake has been subjected to a significant influx of pollutants due to industrial emissions of heavy metals and agricultural emissions, as reported by Liu, He et al. and Peng, Liu et al. [38,39].

The objectives of this study were to determine regional spatial patterns of soil TOC, TN, TP, stoichiometry along the water flow of Yuan River and Li River in WDL, and nutrient/grain size patterns. Additionally, we aimed to assess the status of organic matter and nutrients in the WDL floodplain and identify the main influencing factors. This study is intended to provide valuable insights into the conservation of water bird habitats and wetland ecosystems. This study hypothesized that the (1) concentrations of soil TOC, TN, TP, and their stoichiometric ratio decrease with water flow along tributaries and soil depth and (2) soil TOC, TN, and TP concentrations and grain size distribution are inversely related in the floodplain of WDL.

2. Materials and Methods

2.1. Study Area

West Dongting Lake (30,044 ha) is the main source of drinking and domestic water supply for millions of residents in the Hunan Province, China, and it is also protected under the Yangtze Protection Action and Ecological Civilization by the Chinese government. West Dongting Lake is in the Hanshou County of Hunan Province (111°48′50″–112°16′40″ E; 28°49′50″–9°08′50″ N). The region surrounding WDL has a subtropical monsoon climate, with mean annual precipitation ranging from 1200 to 1350 mm [40]. Each year, the flood season starts with the onset of a monsoon in April with a mean water level of 28.9 m. Water comes mostly from the Yuan River and Li River and causes the water level of WDL to increase to its maximum from July to August with a mean water level of 32.8 m [12,41].

The dominant vegetation on the floodplains of WDLs includes sedges (*Carex* spp.), reeds (*Phragmites australis* (Cav.) Trin. ex Steud), and *Miscanthus* spp. The vegetation in the in-lake delta is dominated by *Rorippa globosa* (Turcz. ex Fisch. and C.A. Mey.) Hayek and *Phalaris arundinácea* (L.) and, along the upstream reaches of the Li River and Yuan River, by *Phragmites australis* (Cav.) Trin. ex Steud, and *Miscanthus lutarioriparius* L. Liu ex S.L. Chen and Renvoize, respectively [37]. *Phragmites australis* (Cav.) Trin. ex Steud and *Miscanthus lutarioriparius* L. Liu ex S.L. Chen and Renvoize in the upstream and midstream sites of the Li River are harvested annually for economic purposes [42].

2.2. Sampling Method

We chose five sampling sites along the Yuan River and Li River waterflow with 10 km intervals in WDL (Yn and Li, respectively). The five sites were distributed across the landscape: upper (YnUp and LiUp), middle (YnMd and LiMd), and lower reaches in the confluence of these two rivers in WDL (Delta) (Table 1; Figure 1), where the water flow is influenced by a big island named Changshan. Soil samples were collected from five plots (20 × 20 m) at each site in January 2017. At the same time, five 1 m × 1 m sample cells were set up at the four corners and the center of each sample plot for plant survey. The plots were placed at 10 m intervals from water to inland on the floodplain. The GPS coordinates of each plot of each sample site were recorded. In the center of each plot, soil samples were collected from depths of 0–10 cm (top), 10–20 cm (middle), and 20–30 cm (deep) using a stainless-steel soil corer (diameter 50 mm). Sufficient soil samples were obtained from every depth (75 total samples: five sites, five plots/replicates, and three depths). Soil samples were put into plastic bags and air-dried at room temperature until they reached a constant weight in the laboratory.

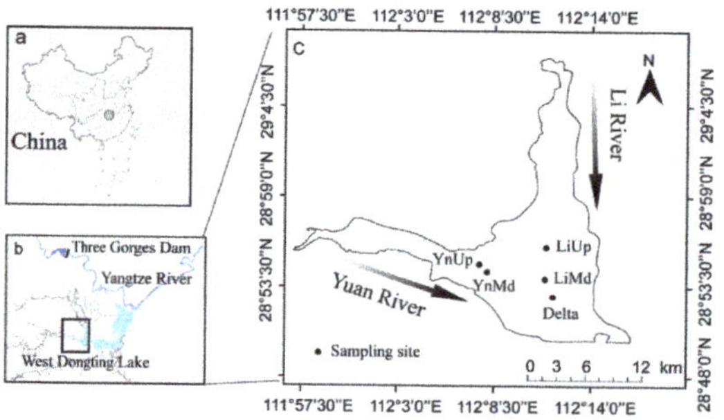

Figure 1. Location of the study area in WDL (**a**) and the location of sampling sites in WDL in Hunan Province (**b**) and China (**c**). Sampling site abbreviations include YnUp (upstream along the Yuan River), YnMd (middle reach of the Yuan River), Delta (Yuan/Li delta), LiUp (upstream Li River), and LiMd (middle reach of the Li River).

Table 1. Locations and abbreviations of the study sites in WDL (Figure 1).

Site Abbreviation	Position	Geographic Location
YnUp	Upstream along Yuan River	28°54′8″ N, 112°8′54.64″ E
YnMd	Middle reach of Yuan River	28°53′35.96″ N, 112°9′22.68″ E
Delta	Yuan and Li Delta	28°51′59.68″ N, 112°13′18.43″ E
LiMd	Middle reach of Li River	28°53′9.36″ N, 112°12′53.04″ E
LiUp	Upstream along Li River	28°55′10.63″ N, 112°12′56.45″ E

2.3. *Chemical Analysis*

The air-dried soil samples were sieved through a 2 mm mesh screen to remove roots and other debris. The homogenized samples were stored in a plastic bag before analysis.

2.3.1. Soil Grain Size Analysis

Hydrogen peroxide and hydrogen chloride were used to remove calcium carbonate from the soil sample. Sodium hexametaphosphate was added to the soil as a dispersant to prepare for testing. The samples were analyzed via a laser diffraction particle size analyser (LDPSA, Microtrac S3500, Microtrac Inc., Orlando, FL, USA) to measure soil grain size [43].

2.3.2. Soil pH Testing

Powdered soil and dd-H_2O were used to test the soil pH of soil samples using a 1:1 soil–water slurry and a pH electrode (ChemTron, PH610, Julabo GmbH, Seelbach, Germany).

2.3.3. Nutrient Analysis

Twenty-five grams of soil samples were ground using a polyvinyl fluoride grinding dish. After grinding, the samples were sieved through a 100-mesh (0.15 mm) sieve to prepare for TOC, TN, and TP analysis. Soil TOC was analyzed using a total organic carbon analyzer (Liqui TOC, Elementar Analysensysteme GmbH, Hanau, Germany). Hydrogen peroxide and hydrogen chloride were used to remove calcium carbonate from the soil sample before TOC analysis. Soil TN was analyzed using the Kjeldahl distillation method [44], and TP was analyzed using the perchloric acid-sulfuric acid digestion-molybdenum antimony anti-colorimetric method [45]. The nutrient concentrations and stoichiometric ratios were all based on mass.

2.3.4. Soil Heavy Metal Analysis

Air-dried soil samples sieved through a 100-mesh sieve were taken to analyze heavy metal concentrations (0.5 g). Every sample (0.5 g) was digested using nitric acid, perchloric acid, and hydrogen fluoride and then heated in a 100 mL polyfluoroethylene crucible with a lid until the samples were dry. The above product was dissolved in 5 mL of 50% nitric acid in a 50 mL volumetric flask. All analysis was performed in the Beijing Forestry University Public Analysis Center. The concentrations of total Hg, Pb, Cd, Ni, Zn, and As were analyzed using inductively coupled plasma mass spectrometry (ICP-MS, optima 8X00, PerkinElmer, Inc,. Waltham, MA, USA). Cu was analyzed using an atomic absorption spectrometer (AAS, SPECTRAA-220, Varian, Palo Alto, CA, USA) [39]. The relevant data have been published in another paper [39].

2.4. *Data Analysis Methods*

2.4.1. Calculation of Soil Grain Size

The grain size was calculated based on Folk [46] and the China National Standard [47]. United States Department of Agriculture textural classification was used to classify soil texture [48]. Soil texture was divided into three types: clay (less than 0.002 mm), silt

(0.002~0.05 mm), and sand (0.05~2 mm). Soil grain size was calculated using the following equations [46]:

$$Mz = (\varphi 16 + \varphi 50 + \varphi 84)/3 \quad (1)$$

where Mz is the mean value of soil grain size; $\varphi 16$, $\varphi 50$, $\varphi 84$ is the grain size of the value corresponding to the 16th, 50th and 84th percentiles on the probability accumulation curve φ in phi. Phi = $-\log_2 D$, where D is the diameter in mm. In this study, soil with larger Mz value means smaller soil grain size.

2.4.2. Calculation of Diversity Indices

The Margalef richness index (S), Pielou evenness index (J), Shannon–Wiener diversity index (H'), and Simpson's diversity index (D) in each plot were calculated using the following equations:

$$S = (N-1)/lnN \quad (2)$$

$$J = H'/lnN \quad (3)$$

$$H' = -\sum_{i=1}^{s} P_i ln P_i \quad (4)$$

$$D = 1 - \sum_{i=1}^{s} P_i^2 \quad (5)$$

In the formula, N is the total number of individuals of all species in the sample plot, s is the number of species in the sample plot, and P_i is the proportion of the number of individuals of plant species i in the total number of individuals (N).

2.4.3. Statistical Analysis

The stoichiometric ratio of each pair of TOC, TN, and TP in the soil samples was calculated. The spatial distribution patterns of soil TOC, TN, and TP as well as the stoichiometric ratio at different sites and soil depths were compared using two-way nested ANOVA (site, depth, and depth [site]). Linear regression was applied to nutrient data vs. the mean value of soil grain size (Mz). The relationships between TOC, TN, and TP were analyzed using general linear regression to indicate whether these nutrients were from the same source type. Redundancy analysis (RDA) was used to explore the relationship between soil TOC, TN, TP, heavy metal concentrations, and plant community diversity. ANOVA and linear fits were calculated in JMP® Version 13.2.0 (SAS Institute Inc., Cary, NC, USA) and RDA was performed using Canoco 5.0 (Biometry, Wageningen, The Netherlands) [49]. Tukey comparisons were used to compare mean differences. The values were graphed using SigmaPlot 12.5 (Systat Software, Inc., San Jose, CA, USA) and GraphPad Prism version 8.0.0 for Windows (GraphPad Software Inc., San Diego, CA, USA).

3. Results

3.1. Spatial Distribution of TOC, TN, and TP from Upstream to Downstream

The mean concentrations of TOC, TN, and TP in the floodplain wetland of WDL were 10.78 ± 0.75, 0.80 ± 0.03, and 0.42 ± 0.02 g·kg^{-1}, respectively. The mean concentrations of TOC, TN, and TP differed in soil depth within the same sample sites (two-way nested ANOVA, $p < 0.0001$, $p = 0.0101$, and $p < 0.0001$; Table 2).

Soil TOC and TP concentrations in the top layer were higher than in the middle and deep layers in YnUp and LiMd sites ($p < 0.05$; Figure 2). Soil TOC concentration in the top layer was higher than in the deep layer in YnMd and LiUp sites ($p < 0.05$; Figure 2). Soil TN concentration in the top layer of YnUp, YnMd, and LiUp was higher than in the deep layer of the Delta site ($p < 0.05$; Figure 2). YnUp and LiMd had higher TOC and TP concentrations than the other sites, especially the Delta site ($p < 0.0001$; Table 2). The Delta site had the lowest concentrations of TN ($p = 0.0009$; Table 2). Soil TP concentrations at all sites differed from each other except at YnUp and LiMd, where TP concentrations were higher than other sites, and group means were not significantly different between YnUp and LiMd ($p < 0.0001$; Table 2).

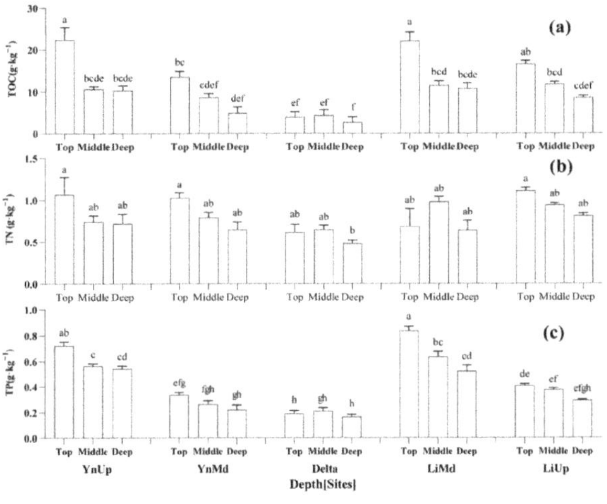

Figure 2. Differences in sites by soil depth (nested effect) on soil nutrient characteristics, including (**a**) TOC, (**b**) TN, and (**c**) TP. The sampling sites include YnUp (upstream along the Yuan River), YnMd (middle reach of the Yuan River), Delta (Yuan/Li Delta), LiMd (middle reach of the Li River), and LiUp (upstream Li River) in WDL, China. Different letters within a soil layer (top, middle, and deep) indicate significant differences among sites. Different letters indicate significant differences between treatments according to the Tukey test at $p < 0.05$.

Table 2. Results of two-way nested ANOVA of the site and soil depth on soil TOC, TN, and TP.

Variables	TOC ($g \cdot kg^{-1}$)				TN ($g \cdot kg^{-1}$)				TP ($g \cdot kg^{-1}$)			
	df	F	*p*	Mean ± S.E.	df	F	*p*	Mean ± S.E.	df	F	*p*	Mean ± S.E.
Site	4	32.6	<0.0001 ***		4	5.4	0.0009 ***		4	171.3	<0.0001 ***	
YnUp				14.43 ± 1.81 a				0.85 ± 0.09 a				0.61 ± 0.03 a
YnMd				8.97 ± 1.18 b				0.82 ± 0.06 a				0.27 ± 0.02 c
Delta				3.53 ± 0.73 c				0.58 ± 0.04 b				0.19 ± 0.01 d
LiMd				14.72 ± 1.62 a				0.77 ± 0.09 ab				0.66 ± 0.04 a
LiUp				12.25 ± 0.95 a				0.96 ± 0.04 a				0.36 ± 0.01 b
Soil depth	2	15.32	<0.0001 ***		2	5.51	0.0059 **		2	3.41	0.0383 *	
Top				15.65 ± 4.59 a				0.90 ± 0.07 a				0.50 ± 0.05 a
Middle				9.32 ± 0.70 b				0.82 ± 0.03 ab				0.41 ± 0.04 ab
Deep				7.38 ± 0.81 b				0.66 ± 0.04 b				0.35 ± 0.03 b
Depth [site]	10	12.39	<0.0001 ***		10	2.63	0.0101 *		10	10.09	<0.0001 ***	

Notes: The sampling sites were located at YnUp (upstream along the Yuan River), YnMd (middle reach of the Yuan River), Delta (Yuan/Li delta), LiUp (upstream Li River), and LiMd (middle reach of the Li River). Soil depths included 0–10 cm (top), 10–20 cm (middle), and 20–30 cm (deep). The significant effects of the variables in the model are indicated by *, **, and *** ($p < 0.05$, $p < 0.01$, and $p < 0.001$, respectively). Different letters indicate significant differences between treatments according to the Tukey test at $p < 0.05$. Depth [site] interaction is presented in Figure 2.

3.2. The Relationship Between Soil TOC, TN, and TP in the Floodplain Wetland

Soil TN and TOC ($r^2 = 0.2639$; $p < 0.001$, Figure 3) as well as soil TP and TOC had positive linear relationships ($r^2 = 0.6530$; $p < 0.001$; Figure 3). The relationship between TN and TP conforms to a linear model, yet the correlation is relatively weak ($r^2 = 0.0867$; $p = 0.01103$; Figure 3).

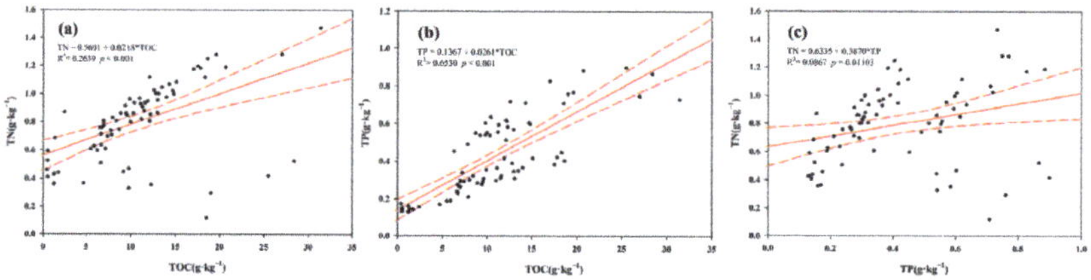

Figure 3. Linear correlation of TN with TOC (**a**), TP with TOC (**b**), and TN with TP (**c**).

3.3. Soil Grain Size and Nutrients

The soil in WDL was primarily silt (Figure 4), mainly ranging in size from 0.00024 mm (12 phi) to 0.0039 mm (8 phi). Sites YnUp, YnMd, LiMd, and LiUp were mainly composed of silt, while Delta was primarily composed of sand (Figure 4).

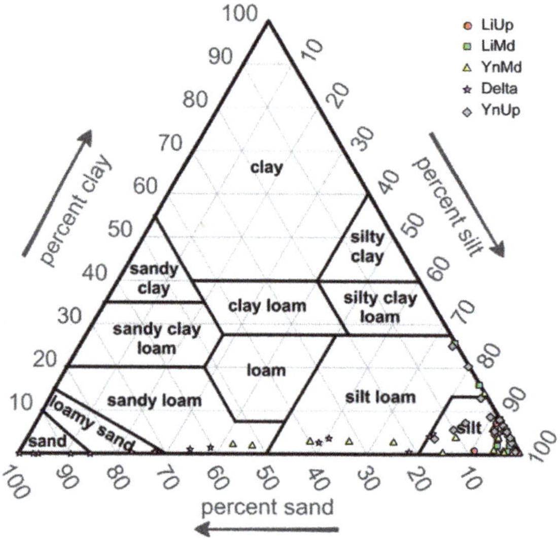

Figure 4. Textural classification of soil in WDL. The red circle represents the soil in LiUp; the green square represents the soil in LiMd; the yellow triangle represents the soil in YnMd; the pink star represents the soil in Delta, and the gray diamond represents the soil in YnUp.

In the general linear regression model, TOC ($r^2 = 0.47$; $p < 0.001$, Figure 5), TN ($r^2 = 0.18$; $p < 0.001$, Figure 5), and TP ($r^2 = 0.49$; $p < 0.001$, Figure 5) were correlated positively with *Mz*, which meant that they were related negatively to soil grain size.

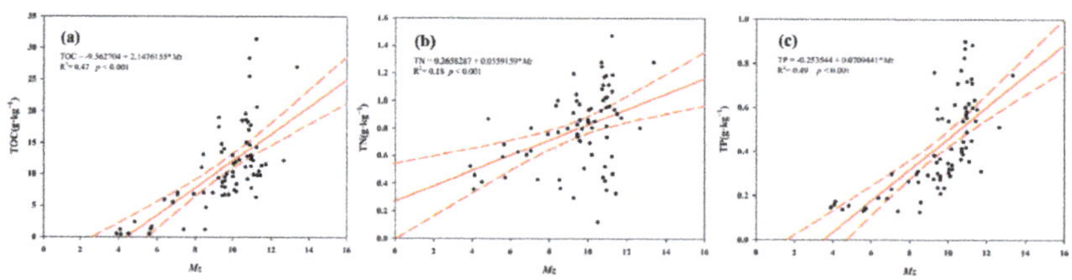

Figure 5. Linear correlation between TOC (**a**), TN (**b**), and TP (**c**) with *Mz* (the mean value of soil grain size; unit was phi, and phi = $-\log_2$D, where D was the diameter in mm).

3.4. Stoichiometric Ratio of TOC, TN, and TP in Soil

The mean C:N and C:P ratios differed in soil depth within the same sample sites (two-way nested ANOVA, $p = 0.0051$, $p < 0.0001$; Table 3). The mean N:P ratio did not vary by depth within the same sample sites ($p > 0.05$; Table 3).

Table 3. Results of two-way nested ANOVA of the site and soil depth on soil C:N, C:P, and N:P.

Variables	C:N				C:P				N:P			
	df	F	p	Mean ± S.E.	df	F	p	Mean ± S.E.	df	F	p	Mean ± S.E.
Site	4	5.24	0.0011 **		4	15.32	<0.0001 ***		4	39	<0.0001 ***	
YnUp				19.34 ± 3.46 ab				22.84 ± 1.99 b				1.39 ± 0.12 b
YnMd				10.08 ± 0.90 b				30.7 ± 2.78 a				3.07 ± 0.07 a
Delta				5.94 ± 1.20 b				16.67 ± 2.89 b				3.21 ± 0.26 a
LiMd				30.22 ± 9.74 a				21.54 ± 1.23 b				1.21 ± 0.14 b
LiUp				12.6 ± 0.57 b				33.67 ± 1.63 a				2.68 ± 0.06 a
Soil depth	2	3.90	0.0245 *		2	8.83	0.0004 ***		2	0.02	0.98	
Top				24.24 ± 6.29 a				31.20 ± 2.10 a				2.29 ± 0.24 a
Middle				11.13 ± 0.77 ab				23.88 ± 1.79 b				2.34 ± 0.18 a
Deep				11.53 ± 1.58 b				20.18 ± 1.75 b				2.30 ± 0.20 a
Depth [site]	10	2.89	0.0051 **		10	4.71	<0.0001 ***		10	0.55	0.85	

Notes: The sampling sites were located at YnUp (upstream along the Yuan River), YnMd (middle reach of the Yuan River), Delta (Yuan/Li Delta), LiUp (upstream Li River), and LiMd (middle reach of the Li River). Variables significantly related to the model are indicated by *, **, and *** ($p < 0.05$, $p < 0.01$, and $p < 0.001$, respectively). Different letters indicate significant differences between treatments according to the Tukey test at $p < 0.05$.

Mean C:N ratio in the top layer was higher than in both the middle and deep layers in LiMd site, with no differences at the other four sites ($p < 0.05$; Figure 6). The top layer of LiMd showed the highest C:N ratio. It was significantly higher than all other layers and sites except for the top layer of YnUp ($p < 0.05$; Figure 6). Mean C:P ratio in the top layer was higher than in the deep layer in YnMd site ($p < 0.05$; Figure 6).

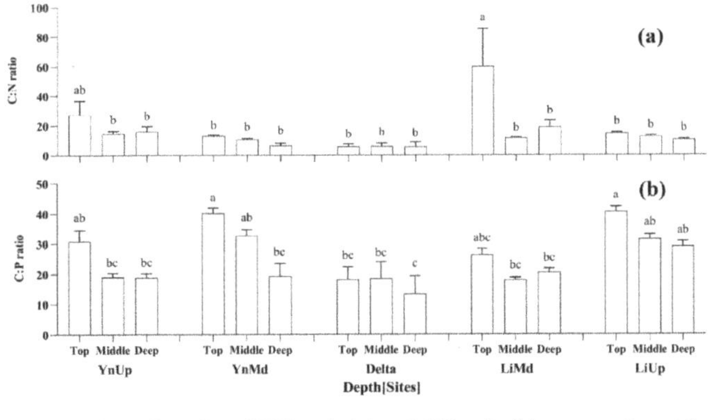

Figure 6. Nested results of C:N ratio (**a**) and C:P ratio (**b**) among sites at three soil depths (0–10 cm, 10–20 cm, and 20–30 cm). The sampling sites include YnUp (upstream along the Yuan River), YnMd (middle reach of the Yuan River), Delta (Yuan/Li Delta), LiMd (middle reach of the Li River), and LiUp (upstream Li River) in WDL, China. Different letters indicate significant differences in mean values at $p < 0.05$.

Mean C:N, C:P, and N:P ratios differed among sites along the water flow direction of the Yuan River and Li River ($p = 0.0011$, $p < 0.0001$, $p < 0.0001$; Table 3). The C:N ratios were higher in YnUp and LiMd; the C:P ratios were higher in YnUp and LiUp, and the N:P ratios were lower in YnUp and LiMd (Table 3). The N:P ratios at the Delta, YnMd, and LiUp sites were higher than those at the YnUp and LiMd sites ($p < 0.0001$; Table 3, Figure 7).

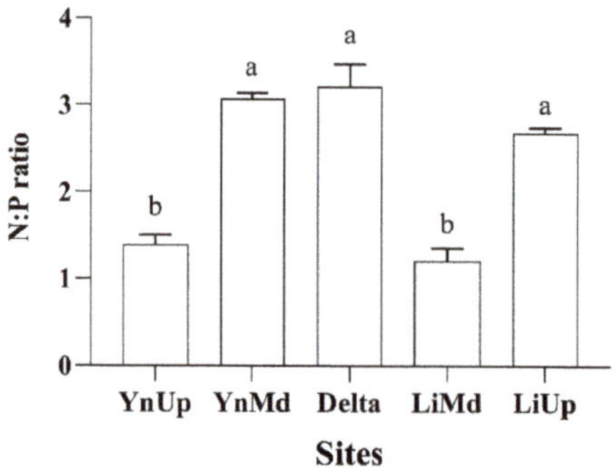

Figure 7. Mean N:P ratios among sites. The sampling sites include YnUp (upstream along the Yuan River), YnMd (middle reach of the Yuan River), Delta (Yuan/Li delta), LiMd (middle reach of the Li River), and LiUp (upstream Li River) in WDL, China. Different letters indicate significant differences in mean values at $p < 0.05$.

3.5. Relationships Between Soil TOC, TN, TP, and Heavy Metal Concentrations and Plant Community Diversity

Relationships between plant diversity and soil TOC, TN, and TP as well as soil heavy metal concentrations were analyzed using RDA (Figure 8). Axes 1 (51.06%) and 2 (10.47%) together explained 61.53% of the total variation. Soil TOC correlated positively with plant diversity ($p = 0.008$, Table 4). Soil Hg was negatively correlated with plant diversity ($p = 0.046$, Table 4). Soil Cd had no direct effect on plant diversity ($p = 0.010$, Table 4).

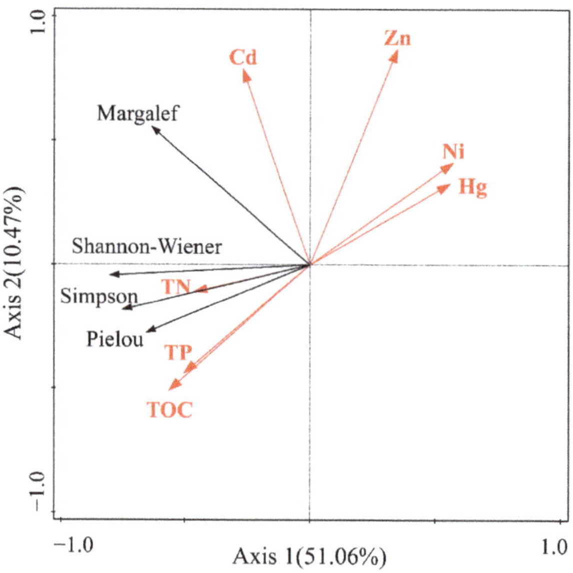

Figure 8. Redundancy analysis (RDA) results between soil TOC, TN, TP, and heavy metal concentrations (red arrows) and plant community diversity (black arrows). Margalef: Margalef richness index, Pielou: Pielou evenness index, Shannon–Wiener: Shannon–Wiener diversity index, and Simpson: Simpson's diversity index.

Table 4. Redundancy analysis (RDA) results of soil TOC, TN, TP, and heavy metal concentrations with species diversity.

Explanatory Variables	Variance Explained (%)	Pseudo-F	p
TOC	18.9	5.4	0.008 **
Cd	17.6	6.1	0.010 *
Hg	7.9	3.3	0.046 *
Ni	6.3	2.4	0.146
TP	5.9	2.7	0.088
Zn	4	1.4	0.256
TN	1.3	0.6	0.522

Notes: The significance is indicated by * and ** ($p < 0.05$, $p < 0.01$, and $p < 0.001$, respectively).

4. Discussion

4.1. Spatial Distribution of TOC, TN, and TP in Soil

Concentrations of TOC, TN, and TP generally decreased downstream along the Yuan River and Li River; however, LiMd had higher TOC and TP concentrations than expected. The higher TOC and TP concentrations of LiMd might be influenced by less *Phragmites australis* and *Miscanthus lutarioriparius* harvesting than LiUp in WDL floodplain [50]. Plants input carbon, nitrogen, and phosphorus into the soil through their litter [51]. Harvesting less means there is more residual plant litter. Flooding may deposit nutrients, especially P, in wetlands via aluminum- or calcium-rich water input [52,53]. The high concentrations of TP in YnUp and LiMd may be related to aluminum-contaminated water input into WDL [39].

The lowest concentrations of TOC, TN, and TP were found in the Delta site at the confluence of the Li River and Yuan River. The Delta site was located at a low elevation in WDL, which is usually the place with the highest sedimentation rate and sediment storage, and sand was the dominant sediment particle type there [23,24,54]. Meanwhile, our results suggested that the soil TOC, TN, and TP concentrations were inversely related to soil grain size, i.e., large sand particles bounded fewer nutrients than smaller soil particles. It could be the main reason why the Delta site has low TOC, TN, and TP concentrations. The study of Pan et al. [55] has also found a similar result. Deltaic sand provides a relatively low amount of TOC, TN, and TP for plants. In contrast, higher concentrations of TOC, TN, and TP were found in the siltier upper and middle reaches of the Li and Yuan Rivers than in the Delta site.

Our results suggested that soil TOC generally decreased with soil depth. Soil C and N concentrations decrease with soil depth and mainly originate from microbial residues and the input of plant root and litter, and soil P shows little variation with the change in soil depth because it is mainly affected by soil parent material [56], which can explain the distribution pattern of TOC with soil depth. The distribution of soil C, N, and P across soil depths is influenced by complex ecological processes, and the unexpected results of TN and TP need further investigation.

4.2. Spatial Distribution of Stoichiometric Ratio of TOC, TN, and TP in Soil

The lower values of C:N and C:P of the Delta site may be related to higher levels of sediment deposition where the water slows. Total nitrogen and TP loads are high, particularly because of upstream anthropogenic inputs of the Li River and Yuan River because of the sediment deposition heterogeneity caused by water flow; the soil C:N:P in Dongting Lake is especially heterogeneous when compared to those in forests, grasslands, and deserts [57]. Usually, in carbon-rich wetland soil, denitrification processes are mainly controlled by soil nitrate concentration and water-filled pore space [58]. Lower soil C:N ratios can accelerate the process of microbial decomposition of organic nitrogen but do not favor carbon accumulation [59]. A lower C:P ratio at the Delta site may be associated with the higher phosphorous availability [32,35,41] and lower mineralization capacity of organic phosphorus [60]. The decrease in C:N ratios and C:P ratios with soil depths in WDL floodplain might be associated with hydrological conditions and oxygen supply, which can limit microbial activity in deeper soils, leading to the slow decomposition of organic matter and changes in ecosystem function [26,61].

4.3. Relationships Between Soil TOC, TN, and TP and Biotic and Abiotic Factors

Our results suggested that the soil TOC, TN, and TP concentrations were inversely related to soil grain size. Soil grain size affects soil TOC, TN, and TP, thereby influencing plant growth conditions. The research conducted by Hamp et al. indicates that soil grain size has a more significant impact on hay productivity and diversity than the effect of atmospheric nitrogen deposition in Ontario's mesic semi-natural grasslands [62].

The dominant plant species at the Delta site were ruderal species such as *Rorippa globosa*, while upstream areas were dominated by tall grasses with extensive root systems such as *Miscanthus lutarioriparius* and *Phragmites australis*. The taller grasses in the floodplains of the tributaries likely promote higher litter production, with higher contributions of organic matter, subsequently increasing soil carbon and nitrogen concentrations in the soil [7,63]. The inputs of aboveground plant residues [64] and the decay of large numbers of dead roots [65] can support high levels of soil organic matter [66].

In the results of RDA, soil TOC was positively related to plant diversity. Previous studies have shown that plant diversity has a positive effect on soil C [67–69]. Higher plant diversity increases plant's nutrient capture and productivity, which have positive feedback effects on soil fertility [68] by influencing plant litter and root litter [69]. Heavy metals such as Hg affect soil TOC in the floodplain through toxic effects on plants, leading to the retardation of growth [70,71].

Soil microbes can influence terrestrial biogeochemistry through the formation and decomposition of organic matter, which consequently influences soil C, N, and P [72,73]. Previous studies have shown that the high content of soil moisture limits soil microbial activities and is unfavorable for decomposition of litter and mineralization, resulting in the accumulation of organic carbon and organic nitrogen [26,36]. Soil oxygen level also affects soil C, N, and P with inundation conditions, which influences the accumulation of wetland soil organic matter [74].

The plant communities could be controlled by the specific stoichiometry characteristics in these floodplains [36,57]. West Dongting Lake had higher C:N ratio and lower C:P and N:P ratio (i.e., C:N, C:P, and N:P are 15.64, 25.08, and 2.31, respectively) than China's mean soil stoichiometry ratio (i.e., C:N, C:P, and N:P are 11.9, 61, and 5.2, respectively) [75]. West Dongting Lake is relatively N limited as indicated by a high C:N ratio (similar to histosols) [76], which suggests higher decomposition rates of organic matter and lower available N. So, the WDL floodplain may tend to be N limited even with very high N deposition, where wet deposition and dry deposition are both high in this region [77]. Wetlands are usually P limited when facing high N deposition [78]. The lower N:P ratio than China's mean soil stoichiometry ratio may be related to the higher P concentration than expected in WDL; P may be input by regional agriculture and/or from other anthropogenic sources or natural sources [77].

We assumed that WDL may possess greater nutrient levels compared to the whole Dongting Lake due to the additional influx of nutrients from Li River and Yuan River into its waters. However, compared with other studies [36], the concentrations of C, N, and P in East Dongting Lake are obviously higher than those in WDL (Table 5) and may be due to intensive anthropogenic activities upstream [41]. The concentrations of TOC, TN, and TP in the coastal Yellow River Delta and Laizhou Bay [63,79] were lower than those in WDL (Table 5). The wetlands in the Ningxia Plain are similar to West Donging Lake and have similar levels of TOC, TN, and TP concentrations [80] (Table 5). These comparisons suggest that WDL may be receiving nutrients from anthropogenic inputs [41]. Reduced anthropogenic activities (agriculture development, watershed urbanization, fishing, and shipping) may be the reason that lower nutrient concentrations is seen in the coastal areas of the Laizhou

Bay and Yellow River than in WDL [41]. Overall, the nutrient patterns in WDL differ from those in other parts of Dongting Lake. This could be due to anthropogenic differences in this sub-watershed, but further studies are needed to confirm the exact reasons.

Table 5. Comparison of soil nutrients between WDL and other wetlands in China.

Wetland Name	Location	Wetland Type	TOC ($g \cdot kg^{-1}$)	TN ($g \cdot kg^{-1}$)	TP ($g \cdot kg^{-1}$)
Ningxia Plain [80]	Middle stream of Yellow River basin.	Lake	8.61	1.05	0.48
Yellow River Delta [79,80]	Middle stream of Yellow River basin. Delta of Yellow River to Bo Sea.	River	6.94	0.56	0.38
		Marsh	9.80	0.96	0.42
		Coastal wetland	5.31	0.46	0.62
Laizhou Bay [63]	West coast of Bo Sea and south of Yellow River Delta.	Coastal wetland	2.93	0.38	0.39
East Dongting Lake [36]	Mid-down stream of Yangtze River basin, middle south of China.	Lake	27.48 ± 2.35	2.78 ± 0.25	1.03 ± 0.06
West Dongting Lake	Mid-down stream of Yangtze River basin, middle south of China.	Lake	10.78	0.80	0.42

5. Conclusions

The distributions of soil TOC, TN, and TP concentrations differed along the Yuan River and Li River in West Dongting Lake. Total organic carbon concentration differed at different sedimentation depths. Soil grain size showed negative effect with soil TOC, TN, and TP concentrations in this region. Plant community diversity correlated positively with soil TOC and negatively with Hg. The results of the study indicate that the wetland of West Dongting Lake is nitrogen limited. Along the direction of river flow, different hydrographic features can affect soil TOC, TN, and TP concentrations. The concentrations of soil TOC, TN, and TP in West Dongting Lake are lower than those in East Dongting Lake which means that the nutrient patterns in WDL differ from those in other parts of Dongting Lake. The present data and analyses can provide valuable insights for the conservation of water bird habitats and wetland ecosystems.

Author Contributions: Conceptualization, T.L.; validation, T.L.; formal analysis, Y.W., M.C., D.P. and L.P.; investigation, J.L., Y.W., D.P. and M.C.; writing—original draft, J.L.; writing—review and editing, J.L., M.C., D.P., L.P., B.A.M. and T.L.; visualization, J.L.; funding acquisition, T.L. All authors have read and agreed to the published version of the manuscript.

Funding: The study was supported by the National Natural Science Foundation of China (No. 42171053).

Data Availability Statement: The data are available from the authors upon reasonable request.

Acknowledgments: We appreciate the anonymous reviewers for the suggestions on revising the manuscript. Any use of trade, firm, or product names is for descriptive purposes only and does not imply endorsement by the US Government.

Conflicts of Interest: Author Mingzhu Chen was employed by Lin Zhiyuan (Beijing) Forestry Engineering Consulting Co., Ltd. All authors declare that the research was conducted in the absence of any commercial or financial relationships that could be construed as a potential conflict of interest.

References

1. Wang, W.; Fraser, J.D.; Chen, J. Wintering Waterbirds in the Middle and Lower Yangtze River Floodplain: Changes in Abundance and Distribution. *Bird Conserv. Int.* **2017**, *27*, 167–186. [CrossRef]
2. Huang, Q.; Sun, Z.; Opp, C.; Lotz, T.; Jiang, J.; Lai, X. Hydrological Drought at Dongting Lake: Its Detection, Characterization, and Challenges Associated with Three Gorges Dam in Central Yangtze, China. *Water Resour. Manag.* **2014**, *28*, 5377–5388. [CrossRef]
3. Pakeman, R.J.; Reid, C.L.; Lennon, J.J.; Kent, M. Possible Interactions between Environmental Factors in Determining Species Optima. *J. Veg. Sci.* **2008**, *19*, 201–208. [CrossRef]

4. Christensen, J.R.; Crumpton, W.G. Wetland Invertebrate Community Responses to Varying Emergent Litter in a Prairie Pothole Emergent Marsh. *Wetlands* **2010**, *30*, 1031–1043. [CrossRef]
5. Post, W.M.; Emanuel, W.R.; Zinke, P.J.; Stangenberger, A.G. Soil Carbon Pools and World Life Zones. *Nature* **1982**, *298*, 156–159. [CrossRef]
6. Thornton, P.E.; Doney, S.C.; Lindsay, K.; Moore, J.K.; Mahowald, N.; Randerson, J.T.; Fung, I.; Lamarque, J.-F.; Feddema, J.J.; Lee, Y.-H. Carbon-Nitrogen Interactions Regulate Climate-Carbon Cycle Feedback: Results from an Atmosphere-Ocean General Circulation Model. *Biogeosciences* **2009**, *6*, 2099–2120. [CrossRef]
7. Bing, H.; Wu, Y.; Zhou, J.; Sun, H.; Luo, J.; Wang, J.; Yu, D. Stoichiometric Variation of Carbon, Nitrogen, and Phosphorus in Soils and Its Implication for Nutrient Limitation in Alpine Ecosystem of Eastern Tibetan Plateau. *J. Soils Sediments* **2016**, *16*, 405–416. [CrossRef]
8. Dodla, S.K.; Wang, J.J.; DeLaune, R.D.; Cook, R.L. Denitrification Potential and Its Relation to Organic Carbon Quality in Three Coastal Wetland Soils. *Sci. Total Environ.* **2008**, *407*, 471–480. [CrossRef]
9. Davidsson, T.E.; Ståhl, M. The Influence of Organic Carbon on Nitrogen Transformations in Five Wetland Soils. *Soil Sci. Soc. Am. J.* **2000**, *64*, 1129–1136. [CrossRef]
10. Maun, M. Adaptations of Plants to Burial in Coastal Sand Dunes. *Can. J. Bot.* **2011**, *76*, 713–738. [CrossRef]
11. Xiao, Y.; Shang, L.; Huang, Z.; Zhang, W.; Xue, Z.; Zhang, Z.; Lu, X. Ecological Stoichiometry Characteristics of Soil Carbon, Nitrogen and Phosphorus in Mountain Swamps of Eastern Jilin Province. *Sci. Geogr. Sin.* **2014**, *34*, 994–1001. [CrossRef]
12. Yuan, Y.; Zeng, G.; Liang, J.; Huang, L.; Hua, S.; Li, F.; Zhu, Y.; Wu, H.; Liu, J.; He, X.; et al. Variation of Water Level in Dongting Lake over a 50-Year Period: Implications for the Impacts of Anthropogenic and Climatic Factors. *J. Hydrol.* **2015**, *525*, 450–456. [CrossRef]
13. Yang, L.; Wang, L.; Yu, D.; Yao, R.; Li, C.; He, Q.; Wang, S.; Wang, L. Four Decades of Wetland Changes in Dongting Lake Using Landsat Observations during 1978–2018. *J. Hydrol.* **2020**, *587*, 124954. [CrossRef]
14. Du, S.; He, C.; Huang, Q.; Shi, P. How Did the Urban Land in Floodplains Distribute and Expand in China from 1992–2015? *Environ. Res. Lett.* **2018**, *13*, 034018. [CrossRef]
15. Bai, J.; Wang, J.; Yan, D.; Gao, H.; Xiao, R.; Shao, H.; Ding, Q. Spatial and Temporal Distributions of Soil Organic Carbon and Total Nitrogen in Two Marsh Wetlands with Different Flooding Frequencies of the Yellow River Delta, China. *CLEAN Soil Air Water* **2012**, *40*, 1137–1144. [CrossRef]
16. Su, Y.; Zhao, H.; Zhao, W.; Zhang, T. Fractal Features of Soil Particle Size Distribution and the Implication for Indicating Desertification. *Geoderma* **2004**, *122*, 43–49. [CrossRef]
17. Tahir, S.; Marschner, P. Clay Addition to Sandy Soil—Influence of Clay Type and Size on Nutrient Availability in Sandy Soils Amended with Residues Differing in C/N Ratio. *Pedosphere* **2017**, *27*, 293–305. [CrossRef]
18. Tyler, S.W.; Wheatcraft, S.W. Fractal Scaling of Soil Particle-Size Distributions: Analysis and Limitations. *Soil Sci. Soc. Am. J.* **1992**, *56*, 362–369. [CrossRef]
19. Flemming, B.W. The Influence of Grain-Size Analysis Methods and Sediment Mixing on Curve Shapes and Textural Parameters: Implications for Sediment Trend Analysis. *Sediment. Geol.* **2007**, *202*, 425–435. [CrossRef]
20. Rajganapathi, V.C.; Jitheshkumar, N.; Sundararajan, M.; Bhat, K.H.; Velusamy, S. Grain Size Analysis and Characterization of Sedimentary Environment along Thiruchendur Coast, Tamilnadu, India. *Arab. J. Geosci.* **2013**, *6*, 4717–4728. [CrossRef]
21. Stribling, J.M.; Cornwell, J.C. Nitrogen, Phosphorus, and Sulfur Dynamics in a Low Salinity Marsh System Dominated by Spartina Alterniflora. *Wetlands* **2001**, *21*, 629–638. [CrossRef]
22. Thoms, M.; Foster, J.M.; Gawne, B. Flood-Plain Sedimentation in a Dryland River: The River Murray, Australia. *IAHS Publ.* **2000**, *263*, 227–236.
23. Walling, D.E.; Owens, P.N.; Leeks, G.J.L. The Role of Channel and Floodplain Storage in the Suspended Sediment Budget of the River Ouse, Yorkshire, UK. *Geomorphology* **1998**, *22*, 225–242. [CrossRef]
24. Steiger, J.; Gurnell, A.M. Spatial Hydrogeomorphological Influences on Sediment and Nutrient Deposition in Riparian Zones: Observations from the Garonne River, France. *Geomorphology* **2003**, *49*, 1–23. [CrossRef]
25. Huang, J.; Niu, L.; Yao, Y.; Qin, H. Influences of the Weather and Climate on Wintering Migratory Bird in Dongting Lake. *Meteorol. Environ. Res.* **2012**, *3*, 29–32, 64.
26. Wang, J.; Bai, J.; Zhao, Q.; Lu, Q.; Xia, Z. Five-Year Changes in Soil Organic Carbon and Total Nitrogen in Coastal Wetlands Affected by Flow-Sediment Regulation in a Chinese Delta. *Sci. Rep.* **2016**, *6*, 21137. [CrossRef]
27. Sánchez-Carrillo, S.; Álvarez-Cobelas, M. Nutrient Dynamics and Eutrophication Patterns in a Semi-Arid Wetland: The Effects of Fluctuating Hydrology. *Water Air Soil Pollut.* **2001**, *131*, 97–118. [CrossRef]
28. Cook, B.J.; Hauer, F.R. Effects of Hydrologic Connectivity on Water Chemistry, Soils, and Vegetation Structure and Function in an Intermontane Depressional Wetland Landscape. *Wetlands* **2007**, *27*, 719–738. [CrossRef]
29. Keller, T.; Håkansson, I. Estimation of Reference Bulk Density from Soil Particle Size Distribution and Soil Organic Matter Content. *Geoderma* **2010**, *154*, 398–406. [CrossRef]
30. Xu, G.; Li, Z.; Li, P. Fractal Features of Soil Particle-Size Distribution and Total Soil Nitrogen Distribution in a Typical Watershed in the Source Area of the Middle Dan River, China. *Catena* **2013**, *101*, 17–23. [CrossRef]
31. Wu, Y.; Wu, Z.; Jiang, S.; Lu, S.; Zhou, N. Elemental Stoichiometry (C, N, P) of Soil in the Wetland Critical Zone of Dongting Lake, China: Understanding Soil C, N and P Status at Greater Depth. *Sustainability* **2022**, *14*, 8337. [CrossRef]

32. Zhang, Z.; Lu, X.; Song, X.; Guo, Y.; Xue, Z. Soil C, N and P Stoichiometry of Deyeuxia Angustifolia and Carex Lasiocarpa Wetlands in Sanjiang Plain, Northeast China. *J. Soils Sediments* **2012**, *12*, 1309–1315. [CrossRef]
33. Ostrowska, A.; Porębska, G. Assessment of the C/N Ratio as an Indicator of the Decomposability of Organic Matter in Forest Soils. *Ecol. Indic.* **2015**, *49*, 104–109. [CrossRef]
34. Wang, S.; Adhikari, K.; Wang, Q.; Jin, X.; Li, H. Role of Environmental Variables in the Spatial Distribution of Soil Carbon (C), Nitrogen (N), and C: N Ratio from the Northeastern Coastal Agroecosystems in China. *Ecol. Indic.* **2018**, *84*, 263–272. [CrossRef]
35. Zhang, Z.-S.; Song, X.-L.; Lu, X.-G.; Xue, Z.-S. Ecological Stoichiometry of Carbon, Nitrogen, and Phosphorus in Estuarine Wetland Soils: Influences of Vegetation Coverage, Plant Communities, Geomorphology, and Seawalls. *J. Soils Sediments* **2013**, *13*, 1043–1051. [CrossRef]
36. Hu, C.; Li, F.; Xie, Y.; Deng, Z.; Chen, X. Soil Carbon, Nitrogen, and Phosphorus Stoichiometry of Three Dominant Plant Communities Distributed along a Small-Scale Elevation Gradient in the East Dongting Lake. *Phys. Chem. Earth Parts A/B/C* **2018**, *103*, 28–34. [CrossRef]
37. Lei, T.; Middleton, B. A U.S.-China EcoPartnership Study of Disturbed Wetland Vegetation in West Dongting Lake, China. *Environ. Prog. Sustain. Energy* **2021**, *40*, e13673. [CrossRef]
38. Liu, M.; He, Y.; Baumann, Z.; Zhang, Q.; Jing, X.; Mason, R.P.; Xie, H.; Shen, H.; Chen, L.; Zhang, W.; et al. The Impact of the Three Gorges Dam on the Fate of Metal Contaminants across the River–Ocean Continuum. *Water Res.* **2020**, *185*, 116295. [CrossRef]
39. Peng, D.; Liu, Z.; Su, X.; Xiao, Y.; Wang, Y.; Middleton, B.A.; Lei, T. Spatial Distribution of Heavy Metals in the West Dongting Lake Floodplain, China. *Environ. Sci. Process. Impacts* **2020**, *22*, 1256–1265. [CrossRef] [PubMed]
40. Jing, L.; Lyu, C.; Yan, Z.; Zuo, A.; Lei, G. Spatio-Temporal Characteristics of the Expansion of Poplar Plantation in West Dongting Lake Wetland, China. *J. Appl. Ecol.* **2016**, *27*, 2039–2047. [CrossRef]
41. Tian, Z.; Zheng, B.; Wang, L.; Li, L.; Wang, X.; Li, H.; Norra, S. Long Term (1997–2014) Spatial and Temporal Variations in Nitrogen in Dongting Lake, China. *PLoS ONE* **2017**, *12*, e0170993. [CrossRef] [PubMed]
42. Dang, N.; Huang, Z.; Li, H. Advances in Biology and Applications of Fiber Plant Triarrhena Lutarioriparia. *J. Trop. Subtrop. Bot.* **2012**, *20*, 418–424. [CrossRef]
43. Wang, D.; Fu, B.; Chen, L.; Wang, Y. Fractal Analysis on Soil Particle Size Distributions under Different Land-Use Types: A Case Study in the Loess Hilly Areas of the Loess Plateau, China. *Shengtai Xuebao/Acta Ecol. Sin.* **2007**, *27*, 3081–3089.
44. *HJ 717-2014*; Soil Quality—Determination of Total Nitrogen—Modified Kjeldahl Method. China Environmental Science Press: Beijing, China, 2014.
45. *HJ 632-2011*; Soil-Determination of Total Phosphorus by Alkali Fusion-Mo-Sb Anti Spectrophotometric Method. China Environmental Science Press: Beijing, China, 2011.
46. Folk, R.L. A Review of Grain-Size Parameters. *Sedimentology* **1966**, *6*, 73–93. [CrossRef]
47. *GB/T 38582-2020*; Specifications for Oceanographic Survey—Part 8: Marine Geology and Geophysics Survey. National Standard of the People's Republic of China: Beijing, China, 2007.
48. United States Department of Agriculture-Natural Resources Conservation Service. *Soil Survey Staff. Keys to Soil Taxonomy*, 12th ed.; United States Department of Agriculture-Natural Resources Conservation Service: Washington, DC, USA, 2014.
49. TerBraak, C.J.F.; Smilauer, P. *Canoco 5, Windows Release (5.12). Software for Multivariate Data Exploration, Testing, and Summarization*; Biometris, Plant Research International: Wageningen, The Netherlands, 2012.
50. Guo, D.; Shi, W.; Qian, F.; Wang, S.; Cai, C. Monitoring the Spatiotemporal Change of Dongting Lake Wetland by Integrating Landsat and MODIS Images, from 2001 to 2020. *Ecol. Inform.* **2022**, *72*, 101848. [CrossRef]
51. Meng, L.; Qu, F.; Bi, X.; Xia, J.; Li, Y.; Wang, X.; Yu, J. Elemental Stoichiometry (C, N, P) of Soil in the Yellow River Delta Nature Reserve: Understanding N and P Status of Soil in the Coastal Estuary. *Sci. Total Environ.* **2021**, *751*, 141737. [CrossRef] [PubMed]
52. Boyer, M.L.H.; Wheeler, B.D. Vegetation Patterns in Spring-Fed Calcareous Fens: Calcite Precipitation and Constraints on Fertility. *J. Ecol.* **1989**, *77*, 597–609. [CrossRef]
53. Richardson, C.J.; Marshall, P.E. Processes Controlling Movement, Storage, and Export of Phosphorus in a Fen Peatland. *Ecol. Monogr.* **1986**, *56*, 279–302. [CrossRef]
54. Scheu, K.R.; Fong, D.A.; Monismith, S.G.; Fringer, O.B. Sediment Transport Dynamics near a River Inflow in a Large Alpine Lake. *Limnol. Oceanogr.* **2015**, *60*, 1195–1211. [CrossRef]
55. Pan, Y.; Zhang, H.; Li, X.; Xie, Y. Effects of Sedimentation on Soil Physical and Chemical Properties and Vegetation Characteristics in Sand Dunes at the Southern Dongting Lake Region, China. *Sci. Rep.* **2016**, *6*, 36300. [CrossRef] [PubMed]
56. Yu, Y.; Chi, Y. Ecological Stoichiometric Characteristics of Soilat Different Depths in a Karst Plateau Mountain Area of China. *Pol. J. Environ. Stud.* **2019**, *29*, 969–978. [CrossRef]
57. Zhang, J.; Li, M.; Xu, L.; Zhu, J.; Dai, G.; He, N. C:N:P Stoichiometry in Terrestrial Ecosystems in China. *Sci. Total Environ.* **2021**, *795*, 148849. [CrossRef]
58. Venterink, H.O.; Davidsson, T.E.; Kiehl, K.; Leonardson, L. Impact of Drying and Re-Wetting on N, P and K Dynamics in a Wetland Soil. *Plant Soil* **2002**, *243*, 119–130. [CrossRef]
59. Srivastava, P.; Pandiaraj, T.; Das, S.; Sinha, S. Characteristics of Soil Organic Carbon, Total Nitrogen and C/N Ratio in Tasar Silkworm Growing Regions of Jharkhand and Bihar States. *Imp. J. Interdiscip. Res.* **2017**, *3*, 426–429.
60. Plante, A.F.; Parton, W.J. 16—The dynamics of soil organic matter and nutrient cycling. In *Soil Microbiology, Ecology and Biochemistry*, 3rd ed.; Paul, E.A., Ed.; Academic Press: San Diego, CA, USA, 2007; pp. 433–467, ISBN 978-0-12-546807-7.

61. Kaushal, S.; Binford, M.W. Relationship between C:N Ratios of Lake Sediments, Organic Matter Sources, and Historical Deforestation in Lake Pleasant, Massachusetts, USA. *J. Paleolimnol.* **1999**, *22*, 439–442. [CrossRef]
62. Hamp, M.; Constant, J.; Grogan, P. Plant Production and Community Structure in a Mesic Semi-Natural Grassland: Moderate Soil Textural Variation Has a Much Stronger Influence than Experimentally Increased Atmospheric Nitrogen Deposition. *Plant Soil* **2024**. [CrossRef]
63. Cao, L.; Song, J.; Li, X.; Yuan, H.; Li, N.; Duan, L.; Wang, Q. Geochemical Characteristics of Soil C, N, P, and Their Stoichiometrical Significance in the Coastal Wetlands of Laizhou Bay, Bohai Sea. *CLEAN Soil Air Water* **2015**, *43*, 260–270. [CrossRef]
64. Li, A.; Deng, W.; Kong, B.; Song, M.; Wenlan, F.; Xiaoning, L.; Lei, G.; Bai, J. A Comparative Analysis on Spatial Patterns and Processes of Three Typical Wetland Ecosystems in 3H Area, China. *Procedia Environ. Sci.* **2010**, *2*, 315–332. [CrossRef]
65. Chen, X.; Li, Y.; Xie, Y.; Deng, Z.; Li, X.; Li, F.; Hou, Z. Trade-off between Allocation to Reproductive Ramets and Rhizome Buds in *Carex Brevicuspis* Populations along a Small-Scale Elevational Gradient. *Sci. Rep.* **2015**, *5*, 12688. [CrossRef]
66. Middleton, B.A. Trends of Litter Decomposition and Soil Organic Matter Stocks across Forested Swamp Environments of the Southeastern US. *PLoS ONE* **2020**, *15*, e0226998. [CrossRef] [PubMed]
67. Zhang, W.; Ren, C.; Deng, J.; Zhao, F.; Yang, G.; Han, X.; Tong, X.; Feng, Y. Plant Functional Composition and Species Diversity Affect Soil C, N, and P during Secondary Succession of Abandoned Farmland on the Loess Plateau. *Ecol. Eng.* **2018**, *122*, 91–99. [CrossRef]
68. Furey, G.N.; Tilman, D. Plant Biodiversity and the Regeneration of Soil Fertility. *Proc. Natl. Acad. Sci. USA* **2021**, *118*, e2111321118. [CrossRef]
69. Lange, M.; Eisenhauer, N.; Sierra, C.A.; Bessler, H.; Engels, C.; Griffiths, R.I.; Mellado-Vázquez, P.G.; Malik, A.A.; Roy, J.; Scheu, S.; et al. Plant Diversity Increases Soil Microbial Activity and Soil Carbon Storage. *Nat. Commun.* **2015**, *6*, 6707. [CrossRef] [PubMed]
70. Gworek, B.; Dmuchowski, W.; Baczewska-Dąbrowska, A.H. Mercury in the Terrestrial Environment: A Review. *Environ. Sci. Eur.* **2020**, *32*, 128. [CrossRef]
71. He, S.; Yang, X.; He, Z.; Baligar, V.C. Morphological and Physiological Responses of Plants to Cadmium Toxicity: A Review. *Pedosphere* **2017**, *27*, 421–438. [CrossRef]
72. Sokol, N.W.; Slessarev, E.; Marschmann, G.L.; Nicolas, A.; Blazewicz, S.J.; Brodie, E.L.; Firestone, M.K.; Foley, M.M.; Hestrin, R.; Hungate, B.A.; et al. Life and Death in the Soil Microbiome: How Ecological Processes Influence Biogeochemistry. *Nat. Rev. Microbiol.* **2022**, *20*, 415–430. [CrossRef] [PubMed]
73. Crowther, T.W.; van den Hoogen, J.; Wan, J.; Mayes, M.A.; Keiser, A.D.; Mo, L.; Averill, C.; Maynard, D.S. The Global Soil Community and Its Influence on Biogeochemistry. *Science* **2019**, *365*, eaav0550. [CrossRef]
74. Steinmuller, H.E.; Chambers, L.G. Characterization of Coastal Wetland Soil Organic Matter: Implications for Wetland Submergence. *Sci. Total Environ.* **2019**, *677*, 648–659. [CrossRef] [PubMed]
75. Tian, H.; Chen, G.; Zhang, C.; Melillo, J.M.; Hall, C.A.S. Pattern and Variation of C:N:P Ratios in China's Soils: A Synthesis of Observational Data. *Biogeochemistry* **2010**, *98*, 139–151. [CrossRef]
76. Wang, S.; Yu, G. Ecological Stoichiometry Characteristics of Ecosystem Carbon, Nitrogen and Phosphorus Elements. *Shengtai Xuebao/Acta Ecol. Sin.* **2008**, *28*, 3937–3947.
77. Lu, C.; Wan, X.; Zeng, W.; Wang, W.; Zhang, Y. Spatio-Temporal Variability of Bulk Nitrogen Deposition in the Dongting Lake Region. *Acta Sci. Circumstantiae* **2018**, *38*, 1137–1146.
78. Bedford, B.L.; Walbridge, M.R.; Aldous, A. Patterns in Nutrient Availability and Plant Diversity of Temperate North American Wetlands. *Ecology* **1999**, *80*, 2151–2169. [CrossRef]
79. Yu, J.; Zhan, C.; Li, Y.; Zhou, D.; Fu, Y.; Chu, X.; Xing, Q.; Han, G.; Wang, G.; Guan, B.; et al. Distribution of Carbon, Nitrogen and Phosphorus in Coastal Wetland Soil Related Land Use in the Modern Yellow River Delta. *Sci. Rep.* **2016**, *6*, 37940. [CrossRef] [PubMed]
80. Bo, X.; Mi, W.; Xu, H.; Zhang, X.; Mi, H.; Song, Y. Contents and ecological stoichiometry characteristics of soil carbon, nitrogen and phosphorus in wetlands of Ningxia plain. *J. Zhejiang Univ. Agric. Life Sci.* **2016**, *42*, 107–118. [CrossRef]

water

Article

Effects of Atmospheric Particulate Matter on Microbial Communities in Wetland Ecosystems

Ying Liu [1,2] and Zhenming Zhang [1,2,*]

[1] School of Ecology and Nature Conservation, Beijing Forestry University, Beijing 100083, China; yingliu33@bjfu.edu.cn
[2] The Key Laboratory of Ecological Protection in the Yellow River Basin of National Forestry and Grassland Administration, Beijing 100083, China
* Correspondence: zhenmingzhang@bjfu.edu.cn; Tel.: +86-13601296335

Abstract: As an important component of urban ecosystems, changes in microbial communities in urban wetland ecosystems have a profound impact on human beings. In this paper, we studied the changes in microbial communities in urban wetland ecosystems (three major interfaces: atmosphere, foliage and water) under the background of atmospheric pollution by high-throughput techniques. The α-diversity of microorganisms at each interface showed that the species richness of the sample communities did not differ significantly at different levels of contamination and it was all at a high level. And the β-diversity showed a significantly larger between-group gap than within-group gap between the samples. The functions predicted a higher metabolic function in water samples and atmospheric samples, and a higher function of microorganisms harmful to humans in the microbial community on the leaf surface. Further analysis of the correlation between atmospheric particulate matter and environmental microorganisms revealed that the atmospheric microbial communities that were strongly negatively correlated with TSP, PM10, PM2.5, and PM1 were *Actinobacteriota*, *Cyanobacteria*, and *Verrucomicrobiota*. Among the microbial communities on the leaf surface, only *Bacteroidota* was strongly positively correlated with total suspended particle (TSP), particles with a diameter of 10 micrometers or less (PM10), particles with a diameter of 2.5 micrometers or less (PM2.5) and particles with a diameter of 1 micrometers or less (PM1). As for the microbial communities in the water column, *Firmicutes*, *Bacteroidota*, *Campilobacterota*, and *Deferribacteres* were strongly and positively correlated with the different particle sizes. There was no significant correlation between the functions of the three interfacial microorganisms and the particle size of the atmospheric particles. This paper studies the structure and function of microbial communities within three interfaces at three pollution levels and explores the resulting changes with the aim of providing directions for monitoring urban wetland ecosystems and for species diversity conservation.

Keywords: atmospheric particulate matter; microbial diversity; wetland ecosystems

1. Introduction

With the rapid social and economic development, environmental problems are becoming more and more prominent, among which the most direct environmental problem faced by human beings is atmospheric pollution [1–3]. Scholars now have a detailed understanding of the composition of atmospheric particulate matter, the sources of matter, and the physicochemical properties of particulate matter [4–7], and clarify its participation in ecological processes such as global energy balance, hydrological cycle, and atmospheric

 https://doi.org/10.3390/w17010066

cycle. At the same time, atmospheric particulate matter also affects other living organisms in the environment [8,9], including environmental microorganisms. The environmental microorganism is a microbial community composed of several populations through symbiosis, mutual benefit, coexistence and competition, which plays an important role in the process of material cycle and energy transformation [10,11].

By conducting numerous studies on the ecological effects between atmospheric particulate matter and environmental microorganisms [12–14], a correlation between atmospheric particulate matter and environmental microorganisms has been found. The relationship between the metabolic characteristics and metabolic functions of atmospheric particulate matter and atmospheric microorganisms has been identified [15–18]. In addition, it has also been shown that atmospheric microorganisms are one of the major sources of human pathogenic pathogens [19]. The effect of atmospheric particulate matter on another environmental microorganism, the interleaf microorganism, has also received some attention [20–25]. Interleaf microorganisms have multiple functions and have important roles in nitrogen fixation, promoting plant growth [26], suppressing pathogenic microorganisms [27,28], improving plant stress resistance [29], and degrading environmental pollutants [30,31], while numerous studies have explored the possible side effects of interleaf microorganisms on leaves [32]. Studies have shown that airborne particles adhering to the surface of plant leaves through sedimentation or adhesion will reduce leaf stomatal conductivity [33] and cause spots on the surface of plant leaves and lead to a decrease in leaf chlorophyll content and photosynthesis. It has also been shown that atmospheric particle deposition also has an effect on water microorganisms [34]. Microorganisms can adapt to changes in environmental factors by indicating changes in environmental factors to reflect the relationship between the microbial community structure in the water body and the surrounding environment [35]. As a decomposer in water bodies, microorganisms in water bodies mainly decompose organic substances such as nitrogen and phosphorus into inorganic substances through nitrification and denitrification of nitrogen, ammonification and decomposition of organic phosphorus, thus regulating the balance of material cycle and energy flow in river ecosystems, reducing the concentration of nutrients in water, promoting the effective use of nitrogen and phosphorus elements by aquatic plants and animals, and playing the role of water purification [36].

However, previous studies are mostly limited to the association between environmental microorganisms and atmospheric particulate matter at a single interface, and although differences in substratum can make great differences in the structure of environmental microbial communities, no attention has been paid to the variability in the effects of environmental changes on environmental microorganisms in the whole ecosystem. Therefore, this study's aim is to expand the study area to focus on a typical urban wetland ecosystem. In most urban wetlands, the atmosphere, forest and water are the three main interfaces in the area, and it is these three components that form the stable ecological structure of urban wetlands. Within the forest–water composite ecosystem, atmospheric particulate matter gradually settles downward, first passing through the atmosphere, then partially adhering to the surface of leaves, and finally reaching the water surface. This study reveals changes in the structure and function of environmental microbial communities within three interfaces under atmospheric deposition conditions: 1. exploring changes in the structure and function of microbial communities in different interface environments; 2. comparing the differences in the microbial community structure at multiple interfaces; 3. analyzing the correlation between atmospheric particulate matter particle size and environmental microorganisms. This study aims to provide basic data on the link between atmospheric deposition and environmental microorganisms within urban artificial wetland ecosystems, and to support further studies to explain the principles of environmental microbial action.

2. Materials and Methods

2.1. Study Area

The experimental area (Figure 1) was selected from the Millennium Watchtower Forest (39°53′35.33″ N, 116°42′48.35″ E) located at Yunfan Road (about 250 m northwest of the Grand Canal Forest Park) in Tongzhou District, Beijing, covering an area of 34.2 ha. The climate of the area is typical of temperate monsoon, with an average annual temperature of 13.8 °C, and abundant sunshine and heat, and it is a typical urban artificial wetland. The rich species diversity in the garden provides high-quality sampling conditions for this experiment.

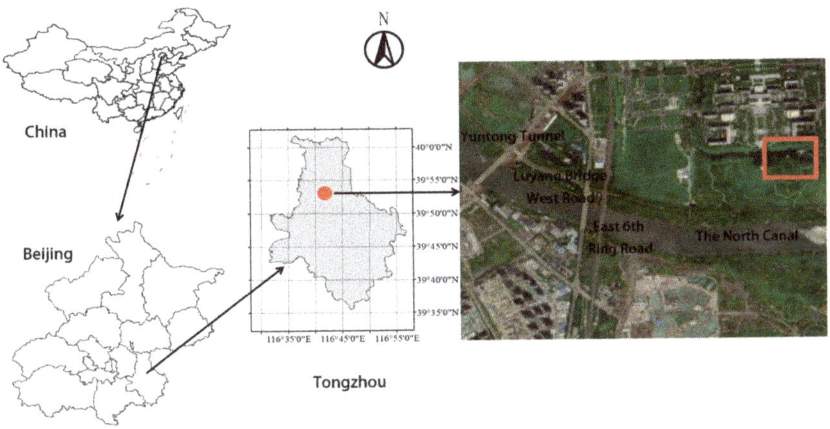

Figure 1. Location of the experimental area.

2.2. Sample Collection

2.2.1. Collection of Atmospheric Particles

Atmospheric particle concentration data include total particle concentration, particles with a diameter of 10 micrometers or less (PM10) concentration, particles with a diameter of 2.5 micrometers or less (PM2.5) concentration and particles with a diameter of 1 micrometers or less (PM1) concentration. The collection instrument was a high-precision handheld dust detector, DUSTMATE, with a setting height of about 1.5 m and a collection time of 12 h. The data collected by DUSTMATE and the air quality index (AQI) found on the website were combined to determine the level of haze pollution (PM2.5, the main pollutant), including good (AQI < 100), light pollution (101 < AQI < 150) and moderate pollution (151 < AQI < 200). Air samples were collected at a single centralized location.

2.2.2. Collection of Atmospheric Microbial Samples

The collection point was set up in the park's weather station flux tower. A model DL-6100D intelligent high-flow-rate particulate sampler was selected for atmospheric microbial collection, and the sampling filter membrane was glass fiber membrane. The sampling airflow rate was controlled at 1.18 m^3/min, the sampling time was 10 h, and the instrument was set to automatically record one set of data every 5 min. A total of three groups were set.

The sampling membrane was autoclaved (121 °C, sterilized for 30 min) before sampling, and then put into a sterile self-sealing bag for slow drying. Alcohol wipe sterilization of experimental equipment before collection of particulate matter was carried out as follows: removal of residual large particles with double-distilled water and 75% alcohol; removal of the glass fiber filter membrane from the sterile self-sealing bag into the collector using sterile forceps; and loading the filter membrane into the filter cartridge, ensuring that the membrane is in the collector and starting the collection. After the samples were collected, the filter membranes were clamped with sterilizing forceps and quickly placed in sterile self-sealing bags, marked with the sample number and placed on dry ice in a constant temperature sterile box for initial storage and transport.

2.2.3. Collection of Microbial Samples on Leaf Surface

The phyllosphere microbial collection point was set up within the oleander forest at the periphery of the flux tower. Medical sterile cotton swabs were used for sampling. A sterile mask and gloves were used and the attached material from the plant leaves was gently wiped. Needles generally wipe 50–100 leaves; the leaves where the height was consistent were wiped, 1.3–1.5 m was deemed appropriate, and a total of three groups were collected. The swab for wiping the adherent was placed in a sterile self-sealing bag, and finally the collected samples were marked with the sample number and placed on dry ice in a constant temperature sterile box for initial storage and transport.

2.2.4. Collection of Microbial Samples in Water

The water microorganism collection point was set in the open water north of the flux tower. Collection was performed using sterile water collection bags. Sterile gloves were used, and surface water of the river was collected. Each sample is 1.5 liters in triplicate. The 0.45 μm and 0.22 μm filter membranes were retained, placed in sterile self-sealing bags, marked with sample numbers, and placed on dry ice in a constant temperature sterile chamber for initial storage and transport. The samples were labeled according to contamination level and sample type (Table 1).

Table 1. Collection sample number.

	Good (AQI < 100)	Light Pollution (101 < AQI < 150)	Moderate Pollution (151 < AQI < 200)
Atmospheric microorganism sample	L.A1	QA1	Z.A1
Water microorganism sample	L.W1	QW1	Z.W1
Phyllosphere microorganism samples	L.L1	QL1	Z.L1

2.3. *Analysis of 16S rRNA Gene Amplicon*

The reagents used in part of this method were obtained from Takara.

2.3.1. Genomic DNA Extraction and PCR Amplification

The genomic DNA of the samples was extracted by the CTAB or SDS methods, and then the purity and concentration of the DNA were detected by agarose gel electrophoresis (Cleaver Scientific, Rugby, Warwickshire, UK). An appropriate amount of the sample DNA was put into a centrifuge tube, and the sample was diluted to 1 ng/μL with sterile water. Using diluted genomic DNA as the template, PCR (Analytik Jena, Jena, Germany) was performed using specific primers with Barcode and efficient high-fidelity enzymes according to the selection of sequencing regions to ensure amplification efficiency and accuracy. Corresponding regions of the primers: 16S V4 primers (515F:GTGCCAGCMGCCGCGGTAA and 806R:GGACTACHVGGGTWTCTAAT): to identify bacterial diversity.

2.3.2. Mixing and Purification of PCR Products

PCR products were detected by electrophoresis with agarose gel of 2% concentration. The qualified PCR products were purified by magnetic beads, and quantified by enzyme labeling (Varioskan LUX, Thermo Scientific, Waltham, MA, USA). The same number of samples were mixed according to the concentration of PCR products. After full mixing, 2% agarose gel electrophoresis was used to detect the PCR products.

2.3.3. Library Construction and Computer Sequencing

Truseq ® DNA PCR-Free Sample Preparation Kit library construction Kit was used for library construction. After Qubit and Q-PCR quantification, the constructed libraries were qualified, and then Novaseq 6000 was used for computer sequencing. There was a certain proportion of Dirty Data in the Raw Data obtained by sequencing. In order to make the results of information analysis more accurate and reliable, the Raw Data were first spliced and filtered to obtain the Clean Data. Then, OTUs' (Operational Taxonomic Units) clustering and species classification were analyzed based on effective data.

2.4. Data Analysis

The data of atmospheric particulate matter were processed simply by Excel. The Illumina Novaseq sequencing platform was used to sequence the gene library to obtain OTUs' clustering. Species annotation analysis was performed using the Mothur method and the SSU rRNA database of Silva138 (http://www.arb-silva.de/) to annotate OTUs' sequences. alpha diversity and beta diversity were analyzed according to the number of OTUs. The calculation of α-diversity indices and β-diversity was based based on Unifrac distances using Qiime software (Version 1.9.1). LEfSe analysis was carried out using LEfSe software (Novo Magic, updated 22 December 2024. https://magic.novogene.com/customer/main#/info-downloadnew/bc9c345abcbe890f8e95cf62c0ebd0b4) to understand significant differences in microbial community data. Functional annotation using Faprotax software (Novo Magic, updated 22 December 2024. https://magic.novogene.com/customer/main#/info-downloadnew/bc9c345abcbe890f8e95cf62c0ebd0b4) was conducted to characterize the functional changes corresponding to the structure of the microbial community. An analysis of the correlation between environmental factors and microbial community structure and function was carried out using the ggcor and ggplot2 packages of the R language software (R 4.1.3, updated 22 December 2024).

3. Results

3.1. Changes in Microbial Community Structure Under Different Pollution Levels

3.1.1. Community Composition and Abundance

Sequences were clustered into OTUs (Operational Taxonomic Units) with 97% identity based on sequencing on the Illumina Nova sequencing platform. Species annotation of OTUs' sequences was carried out against the Silva138 database. Based on the species annotation results, at the phylum level, we found that the dominant ones included mainly *Actinobacteriota, Proteobacteria, Firmicutes*. The dominant species at the genus level were *Bifidobacterium*, unidentified_*Chloroplast*, unidentified_*Blattabacteriaceae*. Dominant species were *Bifidobacterium_animalis, Candidatus_Uzinura_diaspidicola, Parabacteroides_merdae*. The top 10 species with genus-level abundance content were selected for species abundance display (Figure 2), from which the distribution percentages of individual species in different samples and the corresponding changes in abundance can be seen. For example, the proportion of *Machin_Chloroplast* in atmospheric samples is relatively high, and the content of *Machin_Chloroplast* increases first and then decreases with the increase in particulate matter concentration. Especially under the condition of moderate pollution, the abundance of this species decreases dramatically. The percentage of *HGCI-clade* in water samples was relatively high, and the abundance of *HGCI-clade* in water samples changed little under mild pollution conditions, but decreased significantly under moderate pollution conditions. The content of *Parabacteroides* in leaf samples also changed obviously, and the species abundance generally increased with the increase in pollution degree. For *Caproiciproducens*, they all existed in the samples of the three interfaces, and showed a relatively stable trend of increasing species abundance with the increase in pollution level.

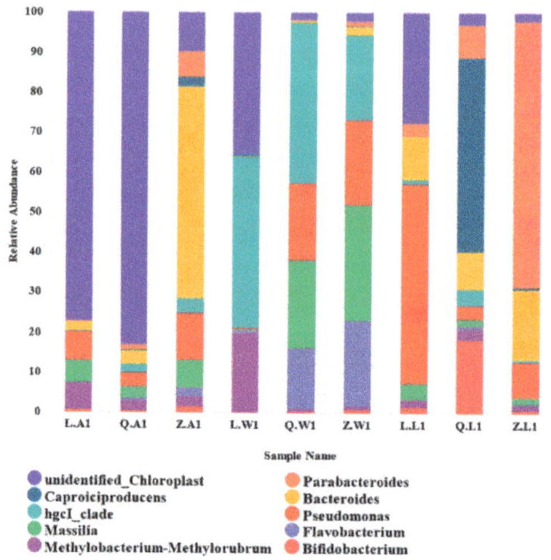

Figure 2. Heatmap of the top 10 species in abundance within each sample (genus level).

Based on the number of OTUs and species annotation, α-diversity and β-diversity were further calculated, and comparisons of differences between groups were carried out, revealing differential characteristics of community structure at different concentrations or environments.

3.1.2. Alpha Diversity of the Community

Usually, alpha diversity refers to the diversity within a community or habitat, and focuses on the species diversity within the community, i.e., species richness and evenness. Table 2 shows the α-diversity indices obtained from the analysis of OTU data after homogenization, where observed_species is the number of OUTs after homogenization. The Simpson's Index of Diversity (1-D) was used in this analysis to characterize the diversity and evenness of species distribution within the community; the higher the value, the greater the diversity of species within the community. The chao1 index is used to indicate community richness, not considering the relative abundance of each species, but only the total number of species; the larger the index, the higher the community richness. The goods_coverage index is used to indicate the sequencing depth index. The table shows that the Simpson, chao1, and goods_coverage indices are at a high level for each interface of the samples, indicating that the collected samples are rich in species diversity and have a high degree of homogeneity. The Simpson's Index of Diversity and the goods_coverage index are above 0.90, which indicates that the species richness of the sample communities is at a high level under different pollution levels.

Table 2. Alpha diversity index of microbial communities.

Group	Observed_Species	Simpson	Chao1	Goods_Coverage
L.A1	3664	0.981	4705.922	0.965
L.L1	2405	0.972	3377.787	0.975
L.W1	1280	0.934	2137.853	0.981
Q.A1	3844	0.929	5302.565	0.956
Q.L1	2698	0.953	3723.494	0.969
Q.W1	1713	0.947	2527.951	0.978
Z.A1	3941	0.967	5349. 213	0.957
Z.L1	2233	0.916	2660.967	0.984
Z.W1	2107	0.973	3235.727	0.972

The alpha diversity of the community can be determined indirectly based on the rarefaction curve, in addition to the magnitude of the index values. When the rarefaction curves tends to be flat, the number of species corresponding to the observed number is the diversity of each sample. The rarefaction curves of each interface at different pollution levels are shown in Figure 3. For both atmospheric and water samples, the community richness showed an increasing trend with the increase in pollution level, and the richness in water samples increased very significantly. For the leaf samples, the community richness showed a trend of increasing and then decreasing with the deepening of the pollution level. Overall, the microbial community diversity was highest in the atmospheric samples, followed by the leaf samples, and the lowest in the water samples among the three interfaces.

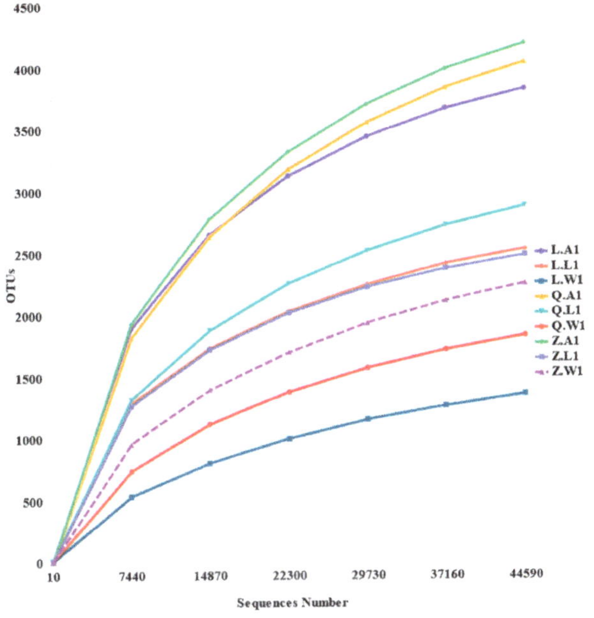

Figure 3. Rarefaction curves of microbial communities at the same interface with different levels of contamination.

3.1.3. Beta Diversity of the Community

Typically, β-diversity refers to the degree of species variation substitution along an environmental gradient, or the dissimilarity in species composition between communities on different habitats due to the environmental gradient. MRPP is a parametric test based on Bray–Curtis distance and was used to analyze whether the differences in microbial community structure between groups were significant. Table 3 shows the results of the MRPP test between two samples. where an A value greater than 0 indicates that between-group differences are greater than within-group differences, and an A value lower than 0 indicates that within-group differences are greater than between-group differences. A smaller Observed-Delta value indicates a smaller intra-group variation, and a larger Expected-Delta value indicates a larger inter-group variation. The data showed a significantly larger between-group gap than within-group gap between samples. It indicates that there is an effect of the change in pollution concentration on the community structure of the samples.

Table 3. MRPP parameters among microbial communities.

Sample	A	Observed-Delta	Expected-Delta
L.A1-Q.A1	0.2605	0.3755	0.5078
Q.A1-Z.A1	0.1843	0.4679	0.5736
L.A1-Z.A1	0.2199	0.4693	0.6015
L.L1-Q.L1	0.1582	0.6207	0.7374
Q.L1-Z.L1	0.07057	0.73	0.7855
L.L1-Z.L1	0.05019	0.6796	0.7155
L.W1-Z.W1	0.278	0.5023	0.6956
L.W1-Q.W1	0.5456	0.2391	0.5261
Q.W1-Z.W1	0.09382	0.5298	0.5847

The Non-Metric Multi-Dimensional Scaling (NMDS, Non-Metric Multi-Dimensional Scaling) statistic is a ranking method applicable to ecological studies. NMDS is a nonlinear model based on Bray–Curtis distances for analysis, reflected as points on a two-dimensional plane based on the species information contained in the sample. It is designed to overcome the shortcomings of linear models and to better reflect the nonlinear structure of ecological data. Applying NMDS analysis, the information about the species contained in the samples is reflected in the form of points on a multidimensional space, while the degree of variation between different samples is reflected by the distance between points, which is able to reflect the inter- and intra-group differences in the samples, etc. Figure 4 shows the NMDS plot made based on the sample data of this experiment. The STRESS value is lower than 0.2, which indicates that the closer the data of the same color are to each other, the less different the communities of samples from the same inner surface will be under the same pollution conditions. The relatively long distance between different colors shows that the differences between sample communities at the same interface under different contamination conditions are much greater than those within the same interface under the same contamination conditions. The variability among the sample communities is shown visually and concretely in the figure.

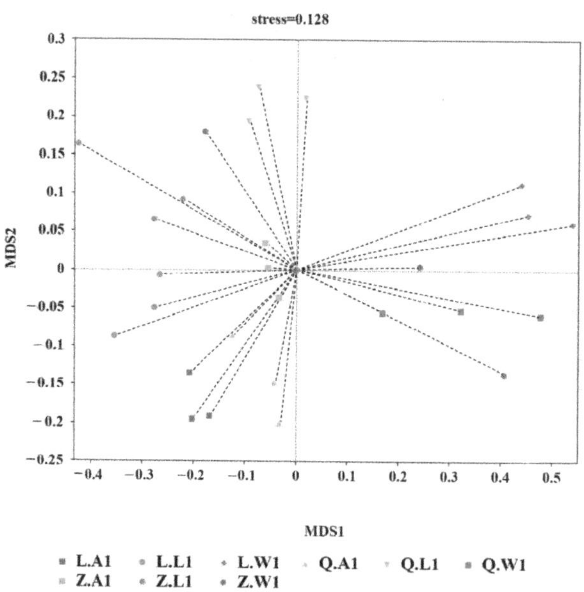

Figure 4. NMDS plots of microbial communities at the same interface with different levels of contamination.

3.1.4. Comparison of Microbial Community Structure at Multiple Interfaces

LefSE analysis is commonly used to analyze species that differ between groups, i.e., to look for biomarkers that are statistically different. The study results include histograms of the distribution of LDA values as well as evolutionary branching plots. One of the bar charts shows the species with significantly different LDA scores greater than the preset value (the preset value of LDA in this paper is 4.0) for each taxonomic species from phylum to family. The longer bar graphs represent the greater degree of influence of significantly different species. The branching diagram represents the taxonomic level from phylum to genus from inside to outside, and the circle in the middle represents the boundary. The size of the diameter of the small circles represents the relative abundance size; species without significant differences are colored the same in yellow, and species with significant differences are colored following the group, i.e., red nodes indicate microbial taxa that play an important role in the red group and blue nodes indicate microbial taxa that play an important role in the blue group. Community names not indicated in the figure are shown in the legend to the right of the image, showing only the phylum to family of differential species. The following three sets of figures show the LefSE analysis of microbial communities in the atmosphere, water bodies and leaf surfaces at different pollution levels (Figure 5).

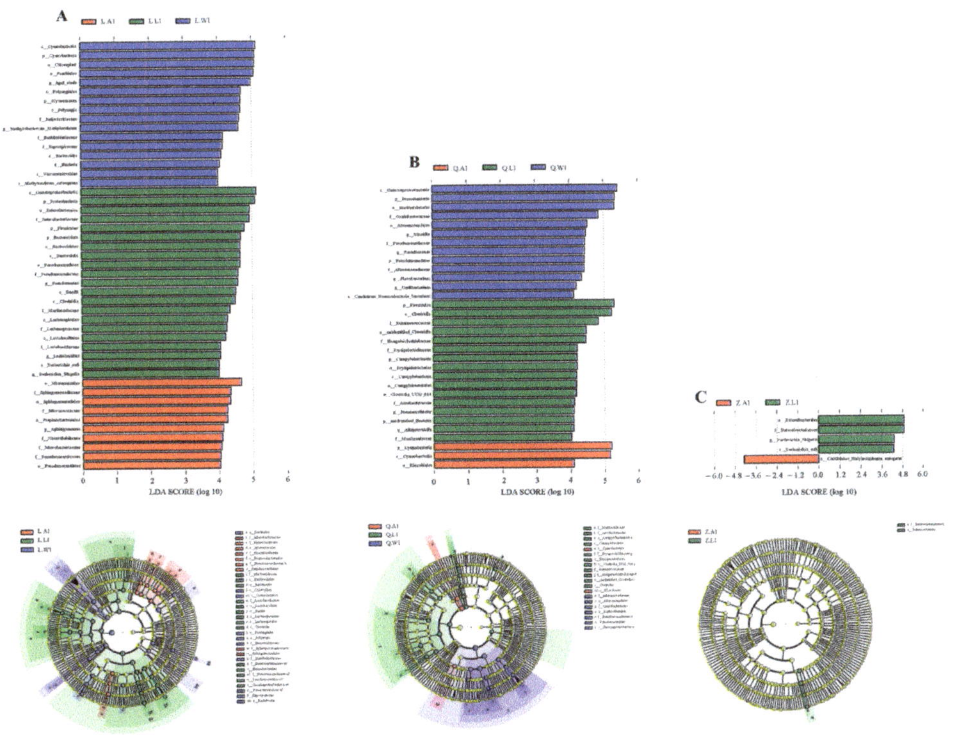

Figure 5. LefSE analysis of microbial communities at different interfaces. Group (**A**,**B**,**C**) plots represent the results of lefse analysis at different levels of contamination, respectively.

As can be seen in Figure 5A, the number of species that differed significantly between interfaces when the pollution level was good was as many as 45. There were 15 species with significant differences in the microbial community of the water body, among which the differences in the *p-Cyanobacteria* had a greater impact. There were 20 species with significant differences in microbial communities on the leaf surface, with a greater impact of differences in *c-Gammaproteobacteria*. There were 10 species with significant differences in atmospheric microbial communities, among which the differences in the *o-Micrococcales* had a greater impact.

As can be seen in Figure 5B, the number of species with significant differences between interfaces amounted to 33 when the pollution level was mild. There were 13 species with significant differences in the microbial community of the water body, among which the differences in the *c-Gammaproteobacteria* had a greater impact. There were 16 species with significant differences in the microbial communities on the leaf surface, with a greater impact of the differences on the *p-Firmicutes*. There were four species with significant differences in atmospheric microbial communities, among which *c-Cyanobacteria* had a greater impact on the differences.

As can be seen in Figure 5C, the species that differed significantly among interfaces at moderate pollution levels were concentrated only in the leaf surface microbial communities. There were four species with significant differences, with a greater effect of differences in the *o-Enterobacterales*. At an LDA score value of 4, there were no species with significant differences between the atmospheric and water column microbial communities.

3.2. Changes in Microbial Community Function

Based on the annotation results of the database, the top 20 functional information values in terms of maximum abundance were selected to generate a functional clustering heat map (Figure 6). The horizontal coordinates in the figure are the sample information and the vertical coordinates are the functional annotation information. The left side of the image is the functional abundance clustering tree, and the top side of the image is the sample clustering tree. The legend indicates the magnitude of functional abundance.

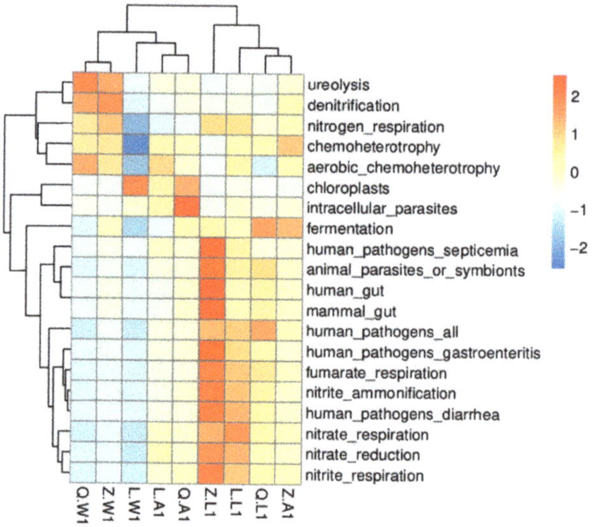

Figure 6. Species functional clustering heat map.

As can be seen from Figure 6, the first eight categories of species functions are mostly concentrated in water samples and atmospheric samples. These eight categories of functions are the main metabolic functions of microorganisms, indicating a strong metabolic function of the microbial community in the water samples. Intracellular_parasites are also highly functional in lightly polluted atmospheres. This also indicates that the metabolic function of the microbial community is relatively active in the water column and that the microorganisms living in that environment are well nourished. As shown in the figure, the microbial functions that can produce pathogenic hazards are mostly concentrated in the leaf surface microbial community. In the case of moderate pollution, the atmospheric microbial community is more pathogenic and dangerous than the aqueous microorganisms. The function of microorganisms harmful to humans in the microbial community of the moderately contaminated leaf surface was the highest among the nine samples. And in the body of water, the pollution situation is always harmful to human microbial function.

3.3. Influence of Atmospheric Particulate Matter Particle Size on Microbial Communities

In this study, we found that the structure and function of multi-interface microbial communities in urban wetland ecosystems changed under different pollution concentrations. Using environmental factor correlation analysis methods, we further analyzed the relationship between the particle size of atmospheric particulate matter and the structure and function of microbial communities within urban wetland ecosystems (Figure 7). The community structure was represented by the top 10 phyla in terms of relative abundance of the samples, and the community function was represented by the six types of biometabolic pathways in KEGG PATHWAY. A, B and C represent the correlation between the structure and function of microbial communities in the atmospheric interface, the leaf interface and the water interface, respectively (the upper part is the community structure, in which the color close to red is a positive correlation, and the color close to green is a negative correlation; the lower part is the community function, and the darker the color, the stronger the correlation), and the particle size of atmospheric particulates.

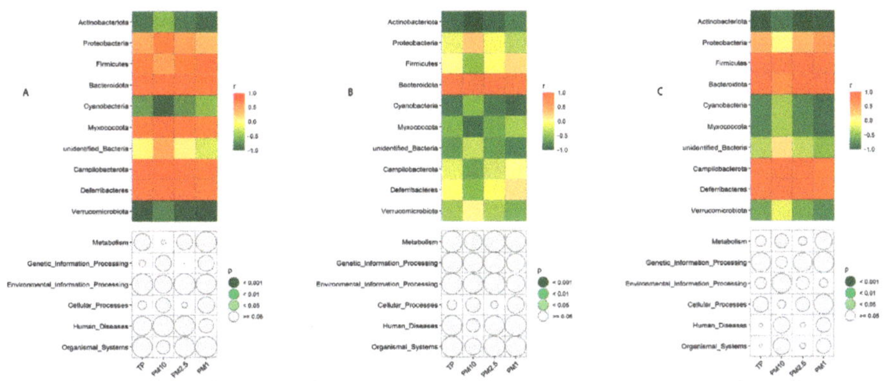

Figure 7. Correlation between particle size of atmospheric particulate matter and the structure and function of microbial communities within urban wetland ecosystems.

From Figure 7A, it can be seen that among the atmospheric microbial communities, *Actinobacteriota*, *Cyanobacteria* and *Verrucomicrobiota* were strongly negatively correlated with TSP, PM10, PM2.5, and PM1. The correlation with TSP, PM10, PM2.5 and PM1 was called weak for unidentified_*Bacteria*. The remaining six phyla were positively correlated with PM10, PM2.5 and PM1. From Figure 7B, it can be seen that the only microbial community on the leaf surface that was strongly positively correlated with TSP, PM10, PM2.5 and PM1 was *Bacteroidota*. Those negatively correlated with TSP, PM10, PM2.5 and PM1 were *Cyanobacteria*, *Myxococcota*. Species of the other seven phyla were weakly correlated with PM10, PM2.5 and PM1 to varying degrees. From Figure 7C, it can be seen that the microbial communities in the water column were strongly and positively correlated with TSP, PM10, PM2.5 and PM1 by *Firmicutes*, *Bacteroidota*, *Campilobacterota*, and *Deferribacteres*. Those negatively correlated with TSP, PM10, PM2.5 and PM1 were *Actinobacteriota*, *Cyanobacteria* and *Myxococcota*. The species of the other three phyla were weakly correlated with PM10, PM2.5 and PM1. There was no significant correlation between the functions of the three interfacial microorganisms and the particle size of atmospheric particles.

4. Discussion

4.1. Changes in Microbial Community Structure Under Different Pollution Levels

Through this study, we found structural and functional changes in the microbial communities at different interfaces under different levels of contamination. The first thing that is affected by atmospheric particulate pollution is the microbial community present in the atmosphere. In this study, we found that the number of OTUs in the atmospheric microbial community gradually increased with the deepening of the pollution level, and the alpha diversity of the community changed, and both the homogeneity and the richness were significantly higher within the community. The β-diversity of the atmospheric microbial community also showed significant variation when samples at different pollution levels were compared two-by-two, with the differences between samples being significantly stronger than those within samples. However, this study assessed the total concentration and viability of microorganisms in the atmosphere and the correlation of microorganisms with meteorological factors and atmospheric pollutants, concluding a weaker correlation between total atmospheric microbial concentrations and air pollutants. This is contrary to the conclusion of [37]. The article suggested that the reasons for such results may be due to differences between studies, different sampling and analysis methods, and inadequate sample sizes. Nevertheless, most studies have shown a strong correlation between microbial communities and atmospheric pollution. Chen et al. in 2020 noted that moderate concentrations of particulate matter (PM) can lead to increased microbial concentrations [38]. This condition occurs mainly because atmospheric particulate matter is the main living environment for atmospheric microorganisms. When the concentration of particulate matter in the atmospheric environment increases, the abundance and homogeneity of microbial communities within the atmosphere must also increase significantly [39,40], but some studies have found that this trend breaks down under conditions of severe pollution. As the accumulated concentration of particulate matter in the air increases, the total number of microorganisms in the air increases first, while under conditions of severe pollution, the haze concentration increases considerably and the number of microorganisms in the air begins to decrease. The reason for this phenomenon is due to the fact that air pollutants reduce the abundance of airborne microorganisms by inhibiting the growth of *bacteria* [41]. It has also been suggested that competition between the toxic effects of chemical contaminants attached to particulate matter and microbial-growth-promoting components may contribute to this result. Studies have shown that bioaerosol concentrations rise for a time after the onset of haze and then decline with the enrichment of chemicals in particulate matter in the later stages of the haze. In conclusion, the effect of atmospheric pollution on the community of atmospheric microorganisms is significant.

Some studies have found that the epiphytic microbiota appears to be more diverse and abundant and has a different composition compared to the endophytic microbiota of plant leaves [32,38,42]. In this study, we explored the changes in the microbial community of leaf surface samples at different levels of contamination. It was found that the number of OTUs showed a trend of increasing and then decreasing with the increase in pollution level, and the α-diversity of microbial communities on the leaf surface gradually decreased. A two-by-two comparison of samples at different contamination levels revealed significant differences in beta diversity between samples. There are very few studies on the effects of climate, especially atmospheric pollution, on microorganisms on the surface of plant leaves. However, the effect of rainfall, a climatic variable, on the structure of microbial communities on the leaf surface has been investigated and confirmed that climatic variables are predictors of foliar bacterial abundance [24]. And the changes in atmospheric pollution that are the focus of this paper do have a corresponding effect on phyllosphere microorganisms. This is not difficult to deduce: the plant leaf surface environment is the main living space for phyllosphere microorganisms, and it has been found that the habitat condition on the leaf surface significantly affects the diversity of phyllosphere bacteria [43]. In turn, the settling and adhesion of atmospheric particulate matter will reduce leaf stomatal conductivity, causing spots to appear on the plant leaf surface, altering the environment on the leaf surface and thus having an impact on the community composition of phyllosphere microorganisms.

Within the water microbial community, the number of OTUs showed a gradual increase with the deepening of pollution, and the α-diversity and β-diversity of the community structure also showed significant changes. The increase in microbial community richness in the water samples was significant, as demonstrated by the dilution curves. The indices characterizing β-diversity also indicate that sample differences at different levels of contamination are much greater than within-group differences within the same sample. Water microorganisms are an essential part of the river ecosystem and are important for the ecological stability of the river as a whole. When a water body is polluted, the structure as well as the function of the microbial community of the water body change significantly. Atta Rasoo et al. found variability in the distribution of microorganisms with tolerance to metal-contaminated areas by studying the composition of microbial communities in Ti-contaminated river sediments [35]. Atmospheric pollution affects water bodies, and thus the degree of atmospheric pollution increases the microbial community in the water body accordingly. A study in 2016 confirmed that an increase in particulate matter within atmospheric aerosols leads to changes in the surface microbial communities of water bodies. The authors observed that phytoplankton biomass showed a trend of first increasing and then decreasing by adding aerosol particulate matter, leading to an increase in nitrate in still water bodies, revealing the effect of aerosol particulate matter on microorganisms in marine surface waters by alleviating nitrogen limitation in organisms. This is consistent with the conclusions drawn in this paper.

We know that each environment is suitable for the growth of relatively fixed microbial species, which is related to the substrates available in the environment for microbial metabolism or growth [44]. Analysis of the number of OTUs revealed that the water and atmospheric environments are relatively stable even under altered pollution levels, but the leaf surface environment is more susceptible to altered pollution. This may be because microbial growth is also related to the temperature, pH, moisture, salinity and other conditions provided by the environment. The environment on the leaf surface is more unstable and easily disturbed compared to the atmosphere and water bodies; so, the microbial community on the leaf surface is more susceptible to alteration. The effect of environmental factors on microorganisms has been the focus of microbial ecology, and researchers have attempted to find a clear relationship between the two through various research methods, but it is still difficult to identify the mechanisms that produce this effect with the current technology. In this study, we focused on the correlation between atmospheric particulate matter particle size and microbial communities at different interfaces under different pollution levels. We found a significant correlation between the top 10 gates of species abundance in different interfaces and atmospheric particulate matter particle size. Previous studies have shown that *Proteobacteria* are highly resistant to low temperatures and widely distributed [44], and the relative abundance of *Proteobacteria* and *Cyanobacteria* in the atmosphere showed a negative correlation, which is consistent with the results of the present study. Overall, *Proteobacteria* had a positive correlation with different particle sizes, while *Cyanobacteria* had a negative correlation with different particle sizes. In addition to this, it has been found that *Actinobacteria* are more likely to adhere to PM10 samples. From this paper, it is evident that the negative correlation between *Actinobacteriota* and PM10 is the weakest in the atmospheric samples, which is consistent with previous studies. *Bacteroidota* and *Firmicutes* belong to the same intestinal microflora, from which *Bacteroidota* are Gram-negative bacteria and *Firmicutes* are Gram-positive bacteria. In some cases, *Bacteroidota* may become pathogenic and synthesize certain active substances [45,46]. There was a strong positive correlation between *Bacteroidota* and particulate matter of different particle sizes in the microbial community on the leaf surface. However, the correlation between *Firmicutes* as well as other gates with different particle sizes was not strong. The harsh environment of the leaf surface may have contributed to this phenomenon of microbial communities on the leaf surface [47,48]. The effect of atmospheric particulate matter of different particle sizes on the microbial community on the leaf surface may be much lower than other environmental factors. On the whole, microorganisms in water bodies and in the atmosphere show the same correlation with particles of different particle size intervals. It has been noted that nitrate nitrogen in water inhibits *Proteobacteria* [49], which could explain the weak positive correlation found in this study between *Proteobacteria* and atmospheric particles of different particle sizes in water samples than between *Proteobacteria* and atmospheric particles of different particle sizes in atmospheric samples. The finding that *Actinobacteria* are more likely to adhere to PM10 samples also holds true for the aqueous microbial community. The negative correlation between *Actinobacteriota* and PM10 in the water samples remained the weakest among the different particle sizes. This also indicates that the microbial community structure in the water samples is similar to that in the atmospheric samples.

4.2. Changes in Microbial Community Functions Under Different Pollution Levels

In addition to structural changes in the microbial community, functional changes are also of interest. In this study, it was found that microbial community functions at different interfaces changed slightly with increasing atmospheric pollution levels, but this change was not significant for the whole microbial community. Also, the effect of environmental factors on microbial community function was not significant. Why did the apparent change in structure not produce a significant functional change? A number of studies have pointed out that the cause of this phenomenon is the synergistic effect within the microbial community. The microbial community itself, as a complex community, contains hundreds of microbial species, and the cultivable microbial species are no more than 1% of all species. In addition to a certain competitive relationship between these microorganisms within a complex community, they also have a synergistic relationship with each other, and this synergy maintains the stability of the microbial community function. So, usually, when no major ecosystem changes occur, the microbial community is able to quickly coordinate its internal functions and maintain stability. This is the main reason why microorganisms have existed on earth for a longer period of time and are more widely distributed compared to humans.

By comparing the microbial community functions at different interfaces, we found that the pathogenicity function of microbial communities on the surface of water bodies was always the weakest, while the pathogenicity function of microbial communities on the surface of leaves was prominent as the pollution level deepened, while *Bacteroidota*, which can be pathogenic bacteria, also formed a strong positive correlation between leaf samples and atmospheric particles of different particle sizes. The reason for this phenomenon may be due to the retention effect of forests on atmospheric pollution itself. Trees trap damaging microorganisms on the surface of their own leaves, resulting in a more pronounced pathogenic function of the microbial community on the leaf surface [50,51] (Lee et al., 2020; Nele et al., 2015). Therefore, for humans, it is more beneficial for the body to move around polluted water than in polluted woods.

Author Contributions: Conceptualization, Y.L. and Z.Z.; methodology, Y.L.; software, Y.L.; validation, Y.L.; formal analysis, Y.L.; investigation, Y.L.; resources, Y.L.; data curation, Y.L.; writing—original draft preparation, Y.L.; writing—review and editing, Z.Z.; visualization, Y.L.; supervision, Z.Z.; project administration, Z.Z.; funding acquisition, Z.Z. All authors have read and agreed to the published version of the manuscript.

Funding: This research was supported by the Science & Technology Fundamental Resources Investigation Program (2022FY100304), Key Program of National Natural Science Foundation of China
(No. 42330705).

Data Availability Statement: All relevant data are contained within this paper.

Conflicts of Interest: We declare that we do not have any commercial or associative interests that represent conflicts of interest in connection with the work submitted.

References

1. Andreo, W.L.; Pinto, J.A.; Pedruzzi, R.; Kumar, P.; Albuquerque, T. Quantifying the impact of particle matter on mortality and hospitalizations in four Brazilian metropolitan areas. *J. Environ. Manag.* **2020**, *270*, 110840. [CrossRef] [PubMed]
2. Mosley, S. Environmental History of Air Pollution and Protection. In *The Basic Environmental History*; Springer International Publishing: Cham, Switzerland, 2014; pp. 143–169.
3. Zhou, H.; Jiang, M.; Huang, Y.; Wang, Q. Directional spatial spillover effects and driving factors of haze pollution in North China Plain. *Resour. Conserv. Recycl.* **2021**, *169*, 105475. [CrossRef]
4. Bari, M.A.; Baumbach, G.; Kuch, B.; Scheffknecht, G. Wood smoke as a source of particle-phase organic compounds in residential areas. *Atmos. Environ.* **2008**, *43*, 4722–4732. [CrossRef]
5. Grantz, D.A.; Garner, J.H.B.; Johnson, D.W. Ecological effects of particulate matter. *Environ. Int.* **2003**, *29*, 213–239. [CrossRef]
6. Thorpe, A.; Harrison, R.M. Sources and properties of non-exhaust particulate matter from road traffic: A review. *Sci. Total Environ.* **2008**, *400*, 270–282. [CrossRef]
7. Widory, D. Nitrogen isotopes: Tracers of origin and processes affecting PM10 in the atmosphere of Paris. *Atmos. Environ.* **2007**, *41*, 2382–2390. [CrossRef]
8. Gao, J.F.; Fan, X.Y.; Li, H.Y.; Pan, K.L. Airborne Bacterial Communities of PM2.5 in Beijing-Tianjin-Hebei Megalopolis, China as Revealed By Illumina MiSeq Sequencing: A Case Study. *Aerosol Air Qual. Res.* **2016**, *17*, 788–798. [CrossRef]
9. Ji, L.; Zhang, Q.; Fu, X.; Zheng, L.; Dong, J.; Wang, J.; Guo, S. Feedback of airborne bacterial consortia to haze pollution with different PM2.5 levels in typical mountainous terrain of Jinan, China. *Sci. Total Environ.* **2019**, *695*, 133912. [CrossRef]
10. Gandolfi, I.; Bertolini, V.; Bestetti, G.; Ambrosini, R.; Innocente, E.; Rampazzo, G.; Papacchini, M.; Franzetti, A. Spatio-temporal variability of airborne bacterial communities and their correlation with particulate matter chemical composition across two urban areas. *Appl. Microbiol. Biotechnol.* **2015**, *99*, 4867–4877. [CrossRef]
11. Smets, W.; Moretti, S.; Denys, S.; Lebeer, S. Airborne bacteria in the atmosphere: Presence, purpose, and potential. *Atmos. Environ.* **2016**, *139*, 214–221. [CrossRef]
12. Sun, Y.; Xu, S.; Zheng, D.; Li, J.; Tian, H.; Wang, Y. Effects of haze pollution on microbial community changes and correlation with chemical components in atmospheric particulate matter. *Sci. Total Environ.* **2018**, *637–638*, 507–516. [CrossRef] [PubMed]
13. Zhang, X.; Hu, B.X.; Ren, H.; Zhang, J. Composition and functional diversity of microbial community across a mangrove-inhabited mudflat as revealed by 16S rDNA gene sequences. *Sci. Total Environ.* **2018**, *633*, 518–528. [CrossRef]
14. Zhong, S.; Zhang, L.; Jiang, X.; Gao, P. Comparison of chemical composition and airborne bacterial community structure in PM2.5 during haze and non-haze days in the winter in Guilin, China. *Sci. Total Environ.* **2019**, *655*, 202–210. [CrossRef]
15. Liu, H.; Zhang, X.; Zhang, H.; Yao, X.; Zhou, M.; Wang, J.; He, Z.; Zhang, H.; Lou, L.; Mao, W.; et al. Effect of air pollution on the total bacteria and pathogenic bacteria in different sizes of particulate matter. *Environ. Pollut.* **2018**, *233*, 483–493. [CrossRef]
16. Lu, R.; Li, Y.; Li, W.; Xie, Z.; Fan, C.; Liu, P.; Deng, S. Bacterial community structure in atmospheric particulate matters of different sizes during the haze days in Xi'an, China. *Sci. Total Environ.* **2018**, *637–638*, 244–252. [CrossRef]
17. Zhen, Q.; Deng, Y.; Wang, Y.; Wang, X.; Zhang, H.; Sun, X. Meteorological factors had more impact on airborne bacterial communities than air pollutants. *Sci. Total Environ.* **2017**, *601–602*, 703–712. [CrossRef]

18. Zhai, Y.; Li, X.; Wang, T.; Wang, B.; Li, C.; Zeng, G. A review on airborne microorganisms in particulate matters: Composition, characteristics and influence factors. *Environ. Int.* **2018**, *113*, 74–90. [CrossRef]
19. Womack, A.M.; Bohannan, B.J.; Green, J.L. Biodiversity and biogeography of the atmosphere. *Philos. Trans. R. Soc. Lond. B Biol. Sci.* **2010**, *365*, 3645–3653. [CrossRef]
20. Berlec, A. Novel techniques and findings in the study of plant microbiota: Search for plant probiotics. *Plant Sci.* **2012**, *193–194*, 96–102. [CrossRef]
21. Leveau, J.H.J. Life of Microbes on Aerial Plant Parts. In *Principles of Plant-Microbe Interactions*; Springer: Cham, Switzerland, 2015; pp. 17–24.
22. Meyer, C.; Bernard, N.; Moskura, M.; Toussaint, M.L.; Denayer, F.; Gilbert, D. Effects of urban particulate deposition on microbial communities living in bryophytes: An experimental study. *Ecotoxicol. Environ. Saf.* **2010**, *73*, 1776–1784. [CrossRef]
23. Newton, A.C.; Gravouil, C.; Fountaine, J.M. Managing the ecology of foliar pathogens: Ecological tolerance in crops. *Ann. Appl. Biol.* **2010**, *157*, 343–359. [CrossRef]
24. Whipps, J.M.; Hand, P.; Pink, D.; Bending, G.D. Phyllosphere microbiology with special reference to diversity and plant genotype. *J. Appl. Microbiol.* **2008**, *105*, 1744–1755. [CrossRef] [PubMed]
25. Yadav, R.K.P.; Papatheodorou, E.M.; Karamanoli, K.; Constantinidou, H.-I.A.; Vokou, D. Abundance and diversity of the phyllosphere bacterial communities of Mediterranean perennial plants that differ in leaf chemistry. *Chemoecology* **2008**, *18*, 217–226. [CrossRef]
26. Okubo, T.; Ikeda, S.; Sasaki, K.; Ohshima, K.; Hattori, M.; Sato, T.; Minamisawa, K. Phylogeny and functions of bacterial communities associated with field-grown rice shoots. *Microbes Environ.* **2014**, *29*, 329–332. [CrossRef]
27. Rastogi, G.; Coaker, G.L.; Leveau, J.H. New insights into the structure and function of phyllosphere microbiota through high-throughput molecular approaches. *FEMS Microbiol. Lett.* **2013**, *348*, 1–10. [CrossRef]
28. Rastogi, G.; Sbodio, A.; Tech, J.J.; Suslow, T.V.; Coaker, G.L.; Leveau, J.H. Leaf microbiota in an agroecosystem: Spatiotemporal variation in bacterial community composition on field-grown lettuce. *ISME J.* **2012**, *6*, 1812–1822. [CrossRef]
29. Rodriguez, R.J.; Henson, J.; Van Volkenburgh, E.; Hoy, M.; Wright, L.; Beckwith, F.; Kim, Y.-O.; Redman, R.S. Stress tolerance in plants via habitat-adapted symbiosis. *ISME J.* **2008**, *2*, 404–416. [CrossRef]
30. Xie, W.Y.; Su, J.Q.; Zhu, Y.G. Arsenite oxidation by the phyllosphere bacterial community associated with Wolffia australiana. *Environ. Sci. Technol.* **2014**, *48*, 9668–9674. [CrossRef]
31. Yutthammo, C.; Thongthammachat, N.; Pinphanichakarn, P.; Luepromchai, E. Diversity and activity of PAH-degrading bacteria in the phyllosphere of ornamental plants. *Microb. Ecol.* **2010**, *59*, 357–368. [CrossRef]
32. Gong, T.; Xin, X.F. Phyllosphere microbiota: Community dynamics and its interaction with plant hosts. *J. Integr. Plant Biol.* **2021**, *63*, 297–304. [CrossRef]
33. Nicola, F.D.; Concha Grana, E.; Lopez Mahia, P.; Lorenzo, S.M.; Prada Rodriguez, D.; Retuerto, R.; Carballeira, A.; Aboal, J.R.; Fernández, J. Evergreen or deciduous trees for capturing PAHs from ambient air? A case study. *Environ. Pollut.* **2017**, *221*, 276–284. [CrossRef] [PubMed]
34. Gao, X.; Gao, W.; Cui, Z.; Han, B.; Yang, P.; Sun, C.; Zheng, L. Biodiversity and degradation potential of oil-degrading bacteria isolated from deep-sea sediments of South Mid-Atlantic Ridge. *Mar. Pollut. Bull.* **2015**, *97*, 373–380. [CrossRef] [PubMed]
35. Rasool, A.; Xiao, T. Response of microbial communities to elevated thallium contamination in river sediments. *Geomicrobiol. J.* **2018**, *35*, 854–868. [CrossRef]
36. Cai, T.-M.; Guan, L.-B.; Chen, L.-W.; Cai, S.; Li, X.-D.; Cui, Z.-L.; Li, S.-P. Enhanced Biological Phosphorus Removal with Pseudomonas putida GM6 from Activated Sludge. *Pedosphere* **2007**, *17*, 624–629. [CrossRef]
37. Chi, M.-C.; Li, C.-S. Fluorochrome in Monitoring Atmospheric Bioaerosols and Correlations with Meteorological Factors and Air Pollutants. *Aerosol Sci. Technol.* **2007**, *41*, 672–678. [CrossRef]
38. Chen, X.; Kumari, D.; Achal, V. A Review on Airborne Microbes: The Characteristics of Sources, Pathogenicity and Geography. *Atmosphere* **2020**, *11*, 919. [CrossRef]
39. Dong, L.; Qi, J.; Shao, C.; Xi, Z.; Chu, C. Concentration and size distribution of total airborne microbes in hazy and foggy weather. *Sci. Total Environ.* **2015**, *541*, 1011–1018. [CrossRef]
40. Haas, D.; Galler, H.; Luxner, J.; Zarfel, G.; Reinthaler, F.F. The concentrations of culturable microorganisms in relation to particulate matter in urban air. *Atmos. Environ.* **2013**, *65*, 215–222. [CrossRef]
41. Guo, Z.; Wang, Z.; Qian, L.; Zhao, Z.; Zhang, C.; Fu, Y.; Li, J.; Zhang, C.; Lu, B.; Qian, J. Biological and chemical compositions of atmospheric particulate matter during hazardous haze days in Beijing. *Environ. Sci. Pollut. Res. Int.* **2018**, *25*, 34540–34549. [CrossRef]
42. Chen, T.; Nomura, K.; Wang, X.; Sohrabi, R.; He, S.Y. A plant genetic network for preventing dysbiosis in the phyllosphere. *Nature* **2020**, *580*, 653–657. [CrossRef]
43. Redford, A.J.; Fierer, N. Bacterial Succession on the Leaf Surface: A Novel System for Studying Successional Dynamics. *Microb. Ecol.* **2009**, *58*, 189–198. [CrossRef] [PubMed]

44. Bowers, R.M.; Lauber, C.L.; Wiedinmyer, C.; Hamady, M.; Fierer, N. Characterization of Airborne Microbial Communities at a High-Elevation Site and Their Potential To Act as Atmospheric Ice Nuclei. *Appl. Environ. Microbiol.* **2009**, *75*, 5121–5130. [CrossRef] [PubMed]
45. Ramachandran, G.; Miguel-Arribas, A.; Abia, D.; Singh, P.K.; Crespo, I.; Gago-Cordoba, C.; Hao, J.A.; Luque-Ortega, J.R.; Alfonso, C.; Wu, L.J.; et al. Discovery of a new family of relaxases in Firmicutes bacteria. *PLoS Genet.* **2017**, *13*, e1006586. [CrossRef] [PubMed]
46. Thomas, F.; Hehemann, J.H.; Rebuffet, E.; Czjzek, M.; Michel, G. Environmental and Gut Bacteroidetes: The Food Connection. *Front. Microbiol.* **2011**, *2*, 93. [CrossRef]
47. Kim, M.; Singh, D.; Lai-Hoe, A.; Go, R.; Adams, J.M. Distinctive Phyllosphere Bacterial Communities in Tropical Trees. *Microb. Ecol.* **2011**, *63*, 674–681. [CrossRef]
48. Lindow, S.E.; Brandl, M.T. Microbiology of the Phyllosphere. *Appl. Environ. Microbiol.* **2003**, *69*, 1875–1883. [CrossRef]
49. Gao, Y.; Lin, S.S.; Li, T.-Y.; Hu, Z. School of Environment, Northeast Normal University. Effects of reclaimed water reuse on soil physicochemical properties and soil bacteria. *J. Northwest Norm. Univ. (Nat. Sci.)* **2019**, *55*.
50. Lee, B.X.Y.; Hadibarata, T.; Yuniarto, A. Phytoremediation Mechanisms in Air Pollution Control: A Review. *Water Air Soil Pollut.* **2020**, *231*, 437. [CrossRef]
51. Weyens, N.; Thijs, S.; Popek, R.; Witters, N.; Przybysz, A.; Espenshade, J.; Gawronska, H.; Vangronsveld, J.; Gawronski, S.W. The Role of Plant–Microbe Interactions and Their Exploitation for Phytoremediation of Air Pollutants. *Int. J. Mol. Sci.* **2015**, *16*, 25576–25604. [CrossRef]

water

MDPI

Article

Estimation of Ecological Water Requirement and Water Replenishment Regulation of the Momoge Wetland

Hongxu Meng [1,2], Xin Zhong [3], Yanfeng Wu [4,*], Xiaojun Peng [5,*], Zhijun Li [1,2,*] and Zhongyuan Wang [6]

1 School of Hydraulic and Electric Power, Heilongjiang University, Harbin 150080, China; menghongxu1024@163.com
2 Cold Region Groundwater Research Institute, Heilongjiang University, Harbin 150080, China
3 School of Architecture and Planning, Shenyang Urban Construction University, Shenyang 110167, China; zhongxinclimate@163.com
4 State Key Laboratory of Black Soils Conservation and Utilization, Northeast Institute of Geography and Agroecology, Chinese Academy of Sciences, Changchun 130102, China
5 Heilongjiang Provincial Hydrology and Water Resources Center, Harbin 150000, China
6 Anguang Town Comprehensive Governance Service Center, Da'an 131302, China; wdy0567@163.com
* Correspondence: wuyanfeng@iga.ac.cn (Y.W.); xiaojunpeng2025@163.com (X.P.); lizhijun78@163.com (Z.L.)

Abstract: Ensuring the ecological water requirements (EWR) suitable for wetlands are upheld is essential for maintaining the stability and health of their ecosystems, a challenge faced by wetlands globally. However, previous studies on EWRs estimation lack a comprehensive consideration of wetlands and still suffer from the problem of rough time scales. Prior studies have predominantly concentrated on its core and buffer zones, neglecting a comprehensive analysis of the wetland's entirety and failing to account for the seasonal variations in EWRs. To fill this gap, we proposed a novel framework for estimating EWRs wetland's entirety to guide the development of dynamic water replenishment strategies. The grey prediction model was used to project the wetland area under different scenarios and designed water replenishment strategies. We then applied this framework in a key wetland conservation area in China, the Momoge Wetland, which is currently facing issues of areal shrinkage and functional degradation due to insufficient EWRs. Our findings indicate that the maximum, optimal, and minimum EWRs for the Momoge Wetland are 24.14×10^8 m^3, 16.65×10^8 m^3, and 10.88×10^8 m^3, respectively. The EWRs during the overwintering, breeding, and flood periods are estimated at 1.92×10^8 m^3, 5.39×10^8 m^3, and 8.73×10^8 m^3, respectively. Based on the predicted wetland areas under different climatic conditions, the necessary water replenishment volumes for the Momoge Wetland under scenarios of dry-dry-dry, dry-dry-normal, dry-normal-dry, and normal-normal-normal are calculated to be 0.70×10^8 m^3, 0.49×10^8 m^3, 0.68×10^8 m^3, and 0.36×10^8 m^3, respectively. In years characterized by drought, the current water replenishment projects are inadequate to meet the wetland's water needs, highlighting the urgent need for the implementation of multi-source water replenishment techniques to enhance the effectiveness of these interventions. The results of this study provide insights for annual and seasonal water replenishment planning and multi-source water management of wetlands with similar problems as the Momoge Wetland. With these new insights, our novel framework not only advances knowledge on the accuracy of wetland ecological water requirement assessment but also provides a scalable solution for global wetland water resource management, helping to improve the ecosystem's adaptability to future climate changes.

Keywords: ecological water requirement; ecological water replenishment; water requirement levels; target system; Momoge Wetland

1. Introduction

Wetlands are among the most crucial ecosystems on Earth, including a suite of critical ecological functions, such as climate regulation, water purification, organic matter production, and water cycle maintenance [1]. However, global climate change and human activities are exacerbating both the rapid deterioration of water quality and the degradation of wetland ecosystems [2,3]. Since 1900, the global wetland ecosystem area has decreased by nearly 50%, a loss rate that significantly surpasses other terrestrial ecosystems [4]. Wetlands, as ecosystems, form through interactions between land and water, and water resources are essential to the formation and evolution of wetlands [5]. Developing countries in Asia are generally facing water scarcity issues. This situation is primarily due to three factors: climate uncertainty, increased water usage resulting from the global temperature rise, and the continuous rise in water requirement driven by population growth and economic development [6]. Climate change has exacerbated the uneven distribution of water resources globally, leading to frequent floods and droughts [7]. At the regional level, disputes over transboundary water resources are becoming increasingly severe, affecting international relations and ecological stability [8]. At the local level, population growth and urbanization cause an imbalance between water resources supply and requirement while also intensifying issues related to water pollution and the overexploitation of water resources [9]. Ensuring the sustainable use of water resources is of great significance for regional stability and ecological security [10]. Consequently, assessing the ecological water requirement of wetlands and their management strategies holds significant theoretical and practical value.

Ecological water requirement is proposed against the backdrop of increasing contradictions between human water use and the water needs of natural ecosystems [11]. The ecological water requirement of wetlands denotes the volumes of water necessary to sustain the structural integrity, ecological functionality, and ecosystem service provision of these ecosystems. This requirement encompasses the water requirement of rivers, wetlands, lakes, and other aquatic ecosystems [12]. Currently, over 200 methodologies are recognized internationally for assessing ecological water requirement [13]. The estimation of wetland ecological water requirement typically employs several methodologies, including the water balance method, the habitat simulated method, and the ecological function method [14]. The ecological function method is grounded in ecosystem ecology theory, taking into account the ecological water requirement of the various functional components within the wetland ecosystem [15]. It is a widely utilized approach for assessing wetland ecological water requirements. Cui et al. [16] delineated wetland ecological water requirements into specific categories: vegetation water requirement, soil water requirement, wildlife habitat water requirement, and replenishment water requirement, highlighting the varying ecological functional requirements of wetlands. Wu et al. [17] integrated hydrological, hydraulic, and ecological habitat methods to construct an ecological water requirement framework for the lower reaches of the Yellow River and used the ecological functional method to calculate the ecological water requirement for wetlands. These studies provide compelling evidence for wetland restoration and protection from an ecological water requirement perspective.

The results of ecological water requirement estimations can also be utilized to assess the need for ecological water replenishment [18]. Ecological water replenishment is the strategic addition of water to ecosystems, designed to counteract structural degradation and functional loss, with the goal of restoring wetlands' natural self-regulation [19]. Assessing the impact of ecological water replenishment on wetland ecosystems is critical for water resource management and ecological restoration [20]. The methods for calculating ecological water replenishment are diverse, usually based on the moisture requirements of the ecosystem and the regional hydrological characteristics, and determined in combination with ecological and hydrological methods [21]. Jiang et al. [22] proposed an integrated method that merges the Tennant method with landscape water requirements for calculating the ecological water requirement of four areas in the eastern part of Xiamen City. This method was then applied, followed by an analysis of water replenishment measures. Zhang et al. [23] utilized the Mike21 model to construct a water environment model for the Baiguishan Reservoir and proposed an optimal ecological water replenishment plan. These studies demonstrate that combining hydrological and ecological methods to calculate ecological water requirements and using environmental models to optimize water replenishment can provide a scientific basis and effective management tools for wetland ecological water replenishment.

The Momoge Wetland, located in the northwest of Jilin Province, exhibits unique wetland landscapes and serves as a crucial stopover for migratory birds along the East Asian flyway. This makes it play a pivotal role in preserving regional ecological balance and biodiversity [24]. Assessing the ecological water requirement of wetlands is crucial for both the rational allocation of water resources within ecosystems and the restoration of their ecological functions. Therefore, evaluating the ecological water requirement of the Momoge Wetland is indispensable for safeguarding its water ecological security [25]. To date, research on the ecological water requirement in the Momoge Wetland has focused on the core and buffer zones [26,27], yet neglecting an integrated evaluation across the entire wetland. Additionally, there exists a gap in studies exploring the ecological water requirement across various seasons, impeding the meeting of practical requirements for ecological water replenishment in the Momoge Wetland. Current research on the dynamic changes and response mechanisms of ecological water requirements in the Momoge Wetland is insufficient, particularly in the context of climate change. Predictive models for ecological water requirement and replenishment strategies in the wetland are in urgent need of optimization. Therefore, this study aims to conduct an in-depth analysis of the trends in ecological water requirement in the Momoge Wetland and to explore the impact of climate change on wetland ecological water requirement. Additionally, it seeks to propose a more scientific model for predicting water replenishment volumes. This is intended to provide a scientific basis for the sustainable management and protection of the wetland.

This research is designed to assess the ecological water requirement and estimate the necessary replenishment volumes in the Momoge Wetland. By doing so, it aims to provide a scientific basis for the restoration, protection, and integrated management of water resources within the wetland ecosystem. To achieve this objective, we first developed a comprehensive index system for calculating ecological water requirements based on the ecological function method. Furthermore, we employed the grey prediction model to project the wetland area under different scenarios and designed water replenishment strategies. We then applied this framework in a key wetland conservation area in China, the Momoge Wetland, which is currently facing issues of areal shrinkage and functional degradation due to insufficient ecological water requirements. The framework integrates the status of the wetland to assess and quantify its ecological water requirement. It employs a grey prediction model to estimate ecological water replenishment volumes under diverse scenarios. This approach thereby establishes a scientific basis for the ecological conservation of wetlands.

2. Data Sources and Research Methods

2.1. Study Area

The Momoge Wetland is situated within Zhenlai County, Baicheng City, Jilin Province, on the western edge of the Songnen Plain. The study area encompasses latitudes from 45°42′ N to 46°18′ N and longitudes from 123°27′ E to 124°04′ E. The total area of the Momoge Wetland is approximately 1442 km^2, and it is the confluence of several rivers, including the Nenjiang and Tao'er Rivers, forming an extensive water network, as shown in Figure 1a. The Momoge Wetland exhibits a temperate continental monsoon climate characterized by distinct seasons [28]. It serves as a critical stopover and breeding site for a multitude of migratory birds. During the spring and autumn migrations, the wetland is home to substantial populations of rare bird species, including the *Grus leucogeranus*, *Grus japonensis*, and *Ciconia boyciana* [29]. The wetland features a diverse array of vegetation types, encompassing herbaceous marshes, shrub marshes, and sparse forest marshes, and is abundant in plant resources like *Phragmites australis (Cav.) Trin. ex Steud.*, *Typha orientalis C. Presl*, and *Carex* L. [30]. In terms of ecological functions, the Momoge Wetland plays a crucial role in climate regulation, water conservation, water quality purification, and biodiversity protection [31]. The Momoge Nature Reserve has a total of 73 villages, with a population of 39,000 people, and the average population density within the wetland is about 27 people per square kilometer [32]. The wetland has an average annual temperature of 4.5 °C and receives an annual sunshine duration of 2924.5 h. It experiences a frost-free period of approximately 160 days [33]. Additionally, the annual precipitation is about 400 mm, predominantly falling between June and September [31]. The annual evaporation rate is approximately 1000 mm.

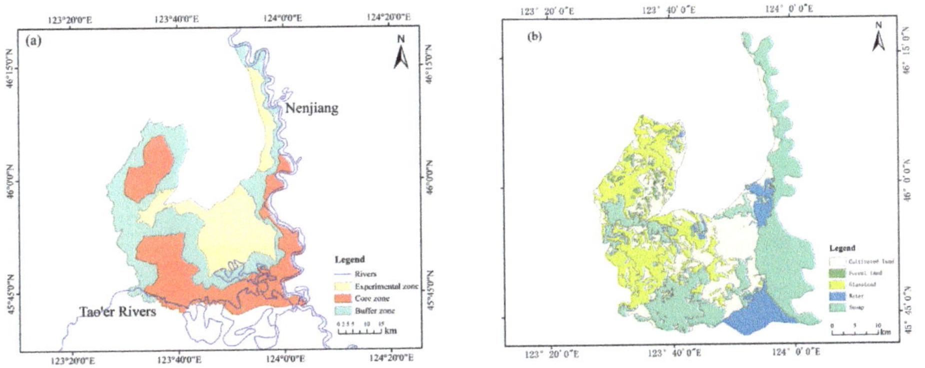

Figure 1. River networks, nature reserve zonation (**a**), and land use types (**b**) in the Momoge Wetland.

The region's topography is characterized by a higher northwest and a lower southeast. It has an overall flat profile with a relative height differential of only 2–10 m. The average elevation is 142 m [34]. The western part is dominated by piedmont sloping terrace plains, while the eastern part comprises large flood plains covered by sporadic dunes [35]. The primary soil types of the reserve include meadow soil, swamp soil, and alluvial soil [28]. The land use types of the Momoge Wetland constitute 23% cultivated land, 2% forest land, 31% grassland, 13% water, and 32% swamp, respectively [36].

2.2. Data Sources and Processing

The data analyzed in this study include soil thickness, evapotranspiration, runoff, rainfall, and the spatial distribution of wetlands. The wetland distribution data are sourced from the global annual wetland dataset at 30 m with a fine classification system from 2000 to 2022 [36]. This dataset features a temporally stable and globally distributed training sample library. It incorporates multi-source remote sensing data, employs locally adaptive modeling techniques, and applies spatiotemporal consistency optimization. This dataset is highly appropriate for examining the spatial patterns and temporal dynamics of wetlands. The soil thickness dataset, with a resolution of the depth-to-bedrock map of China at a spatial resolution of 100 m, was developed by Yan et al. (2020) [37]. Wetland evapotranspiration data are obtained from the ETMonitor Global Actual Evapotranspiration dataset with 1 km resolution [38]. The ETMonitor model estimates daily evapotranspiration at a 1 km pixel scale. It encompasses several components, including vegetation transpiration, evaporation from soils, canopy interception evaporation, evaporation from water surfaces, and snow/ice sublimation. These components are combined to yield the daily evapotranspiration value for each pixel. Meteorological data for the Momoge Wetland and its environs are sourced from the China Meteorological Data Service Centre. Runoff data for the Chaersen and Jiangqiao stations are derived from the China Water Statistical Yearbook [39].

The 2022 spatial distribution data for the Momoge Wetland, encompassing soil thickness and evapotranspiration, were extracted using ArcGIS 10.8. Thereafter, the Momoge Wetland was categorized based on the "Wetland Classification" standard (GB/T24708–2009) [40], as shown in Figure 1b. The Momoge Wetland covers an area of approximately 1442 km^2. In the classification of land uses within the Momoge Wetland, marshlands and extensively vegetated meadows are recognized as natural wetlands. In contrast, paddy fields, which are identified as artificial wetlands, are not included in the ecological water requirement calculations. Statistical analysis reveals the spatial distribution within the Momoge Wetland is divided into three primary categories: water, natural wetlands, and non-vegetated areas. Specifically, the water areas account for 118.45 km^2, the natural wetlands span 757.01 km^2, and the non-vegetated areas cover 391.71 km^2.

2.3. *Ecological Water Requirement Evaluation Index System*

Citing the research of Cui et al. [41], three evaluation indicators were established tailored to the unique characteristics of the Momoge Wetland. These indicators include maintaining the wetland's scale, promoting the conservation of biodiversity, and ensuring the stability of the ecosystem's functions and structure. Maintaining the wetland's scale focuses on preserving and replenishing water losses attributed to evapotranspiration from wetland surfaces and soil, utilizing the evapotranspiration water requirement of wetlands and soil water requirement as evaluation indicators. Promoting the conservation of biodiversity is centered on fulfilling the ecological requirements necessary for the growth and reproduction of wetland species, employing the vegetation water requirement and habitat water requirement as the evaluation indicators for this conservation effort. The stability of the ecosystem's functions and structure prioritizes ensuring adequate groundwater recharge, with the water requirement for groundwater recharge identified as the evaluation indicator for this aspect.

2.4. *Methods for Calculating Ecological Water Requirements*

The ecological function method is based on the calculation of the basic water requirements essential for the diverse functions of the wetland ecosystem. The sustainable ecological water requirement for the wetland ecosystem is derived by aggregating the basic water needs of each functional component [42]. In this study, we selected the entire Momoge Wetland, including the core zone, buffer zone, and experimental zone, as the study area. Based on the structure and function of the Momoge Wetland ecosystem, we utilized the ecological function method to delineate the ecological water requirement into its specific components: evapotranspiration water requirement of wetland, soil water requirement, vegetation water requirement, habitat water requirement, and water requirement for groundwater recharge.

The ecological water requirement calculation model used in this study is:

$$W = W_{P1} + Q_t + W_{P2} + W_P + W_b \quad (1)$$

where W is the ecological water requirement of the Momoge Wetland; W_{P1} is the evapotranspiration water requirement of the wetland; Q is the soil water requirement; W_{P2} is the vegetation water requirement; W_P is the habitat water requirement; and W_b is the water requirement for groundwater recharge.

2.4.1. Evapotranspiration Water Requirement of Wetland

The evapotranspiration water requirement of a wetland encompasses the volume of water lost through evaporation and plant transpiration across a wetland ecosystem. It includes the water requirement for transpiration in vegetated areas and the water requirement for evaporation in non-vegetated areas. Considering the subsequent calculation of the vegetation water requirement in this study, this section concentrates on calculating the evapotranspiration of the non-vegetated area within the wetland. The calculation is given by Equation (2) [43].

$$\frac{dW_{p1}}{dt} = A(t)ET_m \quad (2)$$

where W_{p1} is the evapotranspiration water requirement of the wetland; $A(t)$ is the area of the non-vegetated wetland; ET_m is the evapotranspiration; and t is the time.

2.4.2. Soil Water Requirement

The soil water requirement refers to the optimal moisture content that is essential for sustaining crop growth and enabling various biochemical processes within the soil. Consistent with the research objectives, this analysis utilizes field moisture capacity as a key parameter for determining the soil water requirement. The calculation is given by Equation (3) [15].

$$Q_t = \alpha H_t A_t \quad (3)$$

where Q_t is the soil water requirement; α is the field moisture capacity; H_t is the soil thickness; and A_t is the wetland soil area.

2.4.3. Vegetation Water Requirement

Vegetation water requirement reflects the necessary water quota for the normal growth of plants within a wetland ecosystem. This includes four distinct components: water required for the assimilation process of plants, water retained within plant tissues, water evaporated from the plant surface, and water transpired from the soil. The final two components are jointly known as the ecological water requirement of vegetation, accounting for the majority of water consumption and comprising 99% of the vegetation water requirement [44]. Therefore, vegetation evapotranspiration serves as an estimator for the vegetation water requirement. The calculation is given by Equation (4) [45].

$$W_{p2} = A(t)ET_m \quad (4)$$

where W_{P2} is the vegetation water requirement; $A(t)$ is the area of the vegetated wetland; ET_m is the evapotranspiration; and t is the time.

2.4.4. Habitat Water Requirement

The habitat water requirement encompasses the essential water volume required for fish, birds, and other organisms to inhabit and reproduce in their habitats successfully. In assessing the habitat water requirement, considering the widespread distribution of wetlands, this requirement is typically evaluated based on the proportion of the habitat's water surface area and water depth. The calculation is given by Equation (5) [46].

$$W_p = CHA \tag{5}$$

where W_p is the habitat water requirement; C is the water surface area percentage; H is the suitable water depth; and A is the area of wetland in large-scale areas.

2.4.5. Water Requirement for Groundwater Recharge

Wetlands replenish groundwater through infiltration, which is a process that preserves the stability of the structure and function of wetland ecosystems. The water requirement for groundwater recharge is positively associated with the soil type, hydraulic gradient, and wetland area. The calculation is given by Equation (6) [47].

$$W_b = kIAT \tag{6}$$

where W_b is the water requirement for groundwater recharge; k is the permeability coefficient; I is the hydraulic gradient; and A is the cross-sectional area of the wetland's seepage to groundwater.

2.5. *Predict the Wetland Area*

In this study, we employed a grey prediction model to predict the area of the Momoge Wetland under diverse hydrological conditions. The grey prediction model, integral to grey system theory, is specifically designed for scenarios where data availability is restricted and information is not fully comprehensive [48]. Despite the difficulty in acquiring the temporal sequence of the Momoge Wetland area, the data series for the related influencing factors are more accessible, making the grey prediction model suitable for predicting the wetland area. The inflow from rivers proximate to the Momoge Wetland was identified as a significant influence on the wetland area. This study chose precipitation and the inflows from the Chaersen and Jiangqiao Rivers as the main factors that affect the Momoge Wetland. These factors are important for the grey prediction model. The validity and performance of the developed grey model were evaluated using relative residuals. MATLAB R2022b software was employed for data processing, and the GM (0,3) grey prediction model was employed to conduct an analysis and forecast the area of the Momoge Wetland.

The mechanism and process of grey modeling are as follows [49]:

2.5.1. Data Preprocessing

When conducting grey prediction analysis, we encountered the issue of inconsistent units in the wetland area sequence and the influencing factor sequence. To address this, we normalized these sequences to ensure the comparability of data within the model. Specifically, we normalized the wetland area data and various factors affecting the wetland area (such as precipitation and runoff in the Chaersen and Jiangqiao hydrological stations) to eliminate the impact of units in model analysis. Through this preprocessing, we have improved the predictive accuracy of the model and ensured the reliability of our research findings.

$$\widetilde{x}_i^{(0)}(j) = \frac{x_i{}^{(0)}(j)}{x_i{}^{(0)}(1)} \tag{7}$$

where $\widetilde{x}_i^{(0)}(j)$ represents the normalized value of the i-th sequence at the j-th time point. $x_i^{(0)}(j)$ represents the original value of the i-th sequence at the j-th time point. $x_i^{(0)}(1)$ represents the original value of the i-th sequence at the first time point, commonly used as the basis for normalization.

2.5.2. Generate Cumulative Sequence

To ensure the stability and regularity of the data for the grey prediction model, enhancing forecast accuracy, we generate cumulative sequences from the preprocessed data.

$$x_i^{(1)} = \left\{x_i^{(1)}(1), x_i^{(1)}(2), \ldots x_i^{(1)}(n)\right\} \quad (8)$$

where $x_i^{(1)}(k) = \sum_{j=1}^{k} \widetilde{x}_i^{(0)}(j)$, $k = 1, 2, 3, \ldots, n$.

2.5.3. Construct Grey Predictive Model

The expression for the grey prediction model is:

$$\mathrm{X}_1^{(1)}(\mathrm{k}) = \mathrm{a} + \mathrm{b}_2 x_2^{(1)}(\mathrm{k}) + \mathrm{b}_3 x_3^{(1)}(\mathrm{k}) + \ldots + \mathrm{b}_N x_N^{(1)}(\mathrm{k}) \quad (9)$$

where $\hat{\mathrm{a}} = [\mathrm{a}, \mathrm{b}_2, \mathrm{b}_3, \cdots, \mathrm{b}_\mathrm{n}]^\mathrm{T}$ is the parameter vector of the grey prediction model, a is usually the coefficient in the differential equation, and $b_2, b_3, \ldots, b_n$ are coefficients related to the input variables. $\mathrm{X}_1^{(1)}(k)$ is the predicted value of the model.

2.5.4. Calculate Parameters

List the parameters of the above equations as $\hat{\mathrm{a}} = [\mathrm{a}, \mathrm{b}_2, \mathrm{b}_3, \cdots, \mathrm{b}_\mathrm{n}]^\mathrm{T}$. Use the ordinary least squares to calculate the corresponding parameter estimation vector.

$$\hat{a} = \left(B^T B\right)^{-1} B^T Y_n \quad (10)$$

where $Y_n = \left(x_1^{(1)}(2), x_1^{(1)}(3), \ldots x_1^{(1)}(n)\right)^T$

$$B = \begin{pmatrix} 1 & x_2^{(1)}(2) & \cdots & x_N^{(1)}(2) \\ 1 & x_2^{(1)}(3) & \cdots & x_N^{(1)}(3) \\ \vdots & \vdots & \vdots & \vdots \\ 1 & x_2^{(1)}(n) & \cdots & x_N^{(1)}(n) \end{pmatrix}$$

2.5.5. Restore the Physical Meaning of the Model

In the grey prediction model, the method of cumulative difference is used to transform the accumulated value sequence estimated by the model back to the scale of the original data sequence, thereby restoring the physical significance of the data.

$$\mathrm{X}_1(k+1) = \mathrm{X}_1^{(1)}(k+1) - \mathrm{X}_1^{(1)}(k) \quad (11)$$

Since we normalized the data in Section 2.5.1 to facilitate the model development, it is now necessary to reverse the normalization of the forecasted results to restore their physical significance. To do this, we multiply the estimated values from the model by the first observed value of the original data $x_1^{(0)}(1)$. This process ensures that the predictions are scaled back to the original data's magnitude, maintaining their practical meaning and interpretability for real-world applications.

2.5.6. Validate the Grey Model

The accuracy and reliability of the prediction model can be tested by residual tests and post-checking difference methods [50]. Let the actual observed sequence of wetland area be denoted as ${X_1}^{(0)}$, and the predicted sequence as ${\hat{x_1}}^{(0)}$.

1. Residual tests

The residual sequence, denoted as $\varepsilon(k)$, is the difference between the original sequence and the predicted sequence. The relative error is denoted as Δ_k, and the average relative error is denoted as $\overline{\Delta}$.

$$\Delta_k = \left| \frac{\varepsilon(k)}{{X_1}^{(0)}(k)} \right| \times 100\% = \left| \frac{{X_1}^{(0)}(k) - {\hat{x_1}}^{(0)}(k)}{{x_1}^{(0)}(k)} \right| \times 100\% \quad (12)$$

$$\overline{\Delta} = \frac{1}{n} \sum_{k=1}^{n} \Delta_k, k = 1,2,3,\ldots n \quad (13)$$

When $\overline{\Delta}$ is less than 50%, we consider the grey prediction model to be qualified.

2. Posteriori difference test.

The posteriori difference test, also known as the residual probability test, denotes the variance of the original data sequence as S_1^2, and the variance of the residual sequence as S_2^2.

$$S_1^2 = \frac{1}{n-1} \sum_{K=1}^{n} \left(X_1^{(0)}(k) - \overline{X}_1^0 \right)^2 \quad (14)$$

$$S_2^2 = \frac{1}{n-1} \sum_{K=1}^{n} (\varepsilon(k) - \overline{\varepsilon}(k))^2 \quad (15)$$

where $\overline{X}_1^0 = \frac{1}{n} \sum_{k=1}^{n} X_1^{(0)}(k), k = 1,2,3,\ldots n$

$\overline{\varepsilon}(k) = \frac{1}{n} \sum_{k=1}^{n} \varepsilon(k), k = 1,2,3,\ldots,n$

The variance ratio C and the small error probability P are defined as follows:

$$C = \frac{S_2}{S_1} \quad (16)$$

$$P = \mathrm{Pro}\{|\varepsilon(k) - \overline{\varepsilon}(k)| < 0.6745 S_1\} \quad (17)$$

Posterior difference test criteria as shown in Table 1.

Table 1. The grade of the posterior error.

Model Precision Level	Level 1 (Good)	Level 2 (Qualified)	Level 3 (Marginally Qualified)	Level 4 (Unqualified)
P	$0.95 \le P$	$0.80 \le P < 0.95$	$0.70 \le P < 0.80$	$P < 0.70$
C	$C \le 0.35$	$0.35 < C \le 0.50$	$0.50 < C \le 0.65$	$0.65 < C$

2.5.7. Refine the Grey Model

When the residual test of the constructed grey prediction model shows a large error, a residual correction model can be established to correct the sequence. When the residuals are large, it indicates that the residual sequence still contains a lot of useful information. Replace the actual observed wetland area sequence ${x_1}^{(0)}(j)$ with the residual sequence $\varepsilon(k)$, and then re-establish the prediction model together with the influencing factor sequences $x_i^{(0)}(j)$ $(i > 1)$.

The model data sequence is generally required to be non-negative. For sequences containing negative values, take $b = min\{\varepsilon(k)\}$, and let $\varepsilon'^{(k)} = \varepsilon(k) + |b|$. After making the sequence non-negative, repeat the previous steps with the non-negative sequence to construct the model.

$$\hat{\varepsilon}'(k) = a' + \sum_{i=2}^{n} \mathrm{b}_i \mathrm{x}_i^{(1)}(k), k = 2, 3, \ldots, \mathrm{n} \tag{18}$$

Then, simply reverse the translation to restore the original physical meaning of the model's prediction results.

3. Ecological Water Requirement Levels and Water Allocation During Different Periods of the Year

The ecological water requirement of wetlands displays both threshold-based and seasonal characteristics. This study classifies the ecological water requirement into three specific tiers: maximum, suitable, and minimum. The maximum water requirement indicates the upper limit of water tolerance for a wetland ecosystem; exceeding this threshold may lead to detrimental effects on the wetland's vegetation, wildlife, and soil. The suitable water requirement is the volume of water needed to maintain the structural integrity of the wetland ecosystem, ensure its proper functioning, and achieve its ideal condition. The minimum water requirement represents the critical amount of water essential for maintaining the basic functions of the wetland ecosystem; falling below this level could result in ecosystem degradation. Furthermore, acknowledging the temporal environmental requirements of wetland flora and avifauna, the ecological water requirement is stratified into three specific periods: overwintering period, breeding period, and flood period. During the overwintering period (November to March of the following year), the Momoge Wetland, blanketed by snow, experiences reduced water requirements. The breeding period (April to July) is critical for the proliferation and reproduction of the Momoge Wetland's flora and fauna, thus requiring substantial water supplies. During the flood period (July to October), precipitation increases significantly, vegetation growth is robust, and the wetland's ecological water requirement reaches its zenith.

3.1. Calculation Results of Evapotranspiration Water Requirement of Wetland

The assessment of evapotranspiration water requirement of wetlands is anchored in the proportion of wetland water surface area, serving as a benchmark for differentiating water requirement tiers. Within the Momoge Wetland, the aggregate water surface area, inclusive of both river and non-vegetated regions, defines the maximum water extent, accounting for 35% of the entire natural wetland area. If the water surface area ratio is lower than 13%, wetland vegetation may begin to deteriorate. Accordingly, this study establishes the thresholds for maximum, suitable, and minimum evapotranspiration water requirements of wetlands based on these proportions, corresponding to 0.35, 0.24, and 0.13 of the total natural wetland area, respectively. The evapotranspiration water requirement of a wetland is calculated utilizing Formula (2), and the maximum, suitable, and minimum water requirements for the Momoge Wetland are specified in Table 2. The evapotranspiration water requirements of wetlands during different periods within the year are detailed in Table 3.

Table 2. Ranking of water requirement of wetland evapotranspiration.

Water Requirement Rank	Water Surface Area Percentage	Evapotranspiration Water Requirement of Wetland (Billion m^3)
Maximum	35%	4.33
Suitable	24%	3.06
Minimum	13%	1.64

Table 3. Ranking of ecological water requirements during different periods within the year in the Momoge Wetland (billion m^3).

Periods	Water Requirement Rank	Evapotranspiration Water Requirement of Wetland	Soil Water Requirement	Vegetation Water Requirement	Habitat Water Requirement	Water Requirement for Groundwater Recharge
Overwintering Period	Maximum	0.56	0.14	0.32	1.73	0.08
	Suitable	0.37	0.11	0.31	1.08	0.05
	Minimum	0.20	0.08	0.28	0.65	0.03
Breeding Period	Maximum	2.00	0.74	1.09	3.46	0.44
	Suitable	1.36	0.57	1.03	2.16	0.27
	Minimum	0.72	0.41	0.96	1.30	0.16
Flood Period	Maximum	1.97	2.64	1.06	5.77	1.56
	Suitable	1.31	2.05	1.01	3.61	0.98
	Minimum	0.72	1.46	0.94	2.17	0.59

3.2. Calculation Results of Soil Water Requirement

According to the existing literature [37], it reveals that the soil thickness in the Momoge Wetland is 0.57 m, with the wetland soil area encompassing the natural wetland area of 757 km^2. The soil water requirement is categorized into distinct tiers based on field moisture capacity. When assessing the soil water requirements of the Momoge Wetland, it is considered that the main soil types in the area are meadow soil and swamp soil, with swamp soil accounting for 70% of the total wetland soil area. Therefore, to more accurately reflect the soil moisture needs of natural wetlands, this study selects swamp soil as the primary soil type and uses its field capacity as the key indicator for assessing soil water requirements. When field moisture capacity exceeds 80%, plants reach their peak water tolerance, indicating the maximum soil water requirement. At the 55% mark of field moisture capacity, plants achieve optimal hydration for growth, which is the threshold for the suitable ecological soil water requirement. When field moisture capacity drops below 30%, plants begin to wilt due to soil desiccation, signifying the minimum soil water requirement. Soil water requirements may also be classified based on the proportion of the wetland's average monthly runoff. Monthly runoff data from the Zhenlai station (2000–2014) show that soil water requirement ratios are 0.04 for the overwintering period (November to March), 0.21 for the breeding period (April to June), and 0.75 for the flood period (July to October). The soil water requirement is calculated utilizing Formula (3), and the maximum, suitable, and minimum soil water requirements for the Momoge Wetland are specified in Table 4. The soil water requirements during different periods within the year are detailed in Table 3.

Table 4. Ranking of water requirement of wetland soil.

Water Requirement Rank	Field Moisture Capacity	Soil Water Requirement (Billion m^3)
Maximum	80%	3.52
Suitable	55%	2.73
Minimum	30%	1.94

3.3. Calculation Results of Vegetation Water Requirement

The vegetation water requirement is categorized into specific levels based on both average and potential evapotranspiration rates. The average annual transpiration of the Momoge Wetland is 887 mm, establishing it as the minimum threshold for the ecological water requirement of vegetation. The potential evapotranspiration, at 1000 mm, represents the maximum ecological water requirement for vegetation. This study delineates the maximum, suitable, and minimum water requirements for vegetation corresponding to average annual transpiration levels of 1000 mm, 950 mm, and 887 mm, respectively. Based on the proportion of monthly evapotranspiration within the Momoge Wetland region, the corresponding water requirement ratios for vegetation are 0.44 for the overwintering period (November to March of the following year), 0.43 for the breeding period (April to June), and 0.13 for the flood period (July to October). The vegetation water requirement is calculated utilizing Formula (4), and the maximum, suitable, and minimum water requirements for the Momoge Wetland are specified in Table 5. The vegetation water requirements during different periods within the year are detailed in Table 3.

Table 5. Ranking of water requirement of wetland vegetation.

Water Requirement Rank	Average Evapotranspiration (mm)	Vegetation Water Requirement (Billion m^3)
Maximum	1000	2.47
Suitable	950	2.35
Minimum	887	2.19

3.4. Calculation Results of Habitat Water Requirement

The habitat water requirement is categorized into specific levels based on the percentage of wetland water surface area. This study defines the maximum, suitable, and minimum water requirements for biological habitats corresponding to water surface area percentages of 80%, 50%, and 30%, respectively. Considering the wetland's rich biodiversity, the habitat water requirement is determined using the key species indicator method [51]. The representative indicator species were chosen based on multiple factors. These factors include the biodiversity and ecological status of the study area. Additionally, the changes in water bodies and the characteristics of topography within the wetland were also considered. The distribution and quantity of aquatic plants are closely related to habitat changes. Selecting dominant populations with strong vitality and distinct water requirements as indicator species is an effective means of guiding the establishment of ecological flow. Herein, Phragmites australis—known colloquially as reed—acts as the reference species for assessing the habitat water requirement. The necessary water depth varies with the seasonal hydric requirements of Phragmites. This study calculates the habitat water requirement based on specified water depths: 15 cm for the overwintering period (November to March of the following year), 30 cm for the breeding period (April to June), and 50 cm for the flood period (July to October). The habitat water requirement is calculated utilizing Formula (5), and the maximum, suitable, and minimum water requirements for the Momoge Wetland are specified in Table 6. The habitat water requirements during different periods within the year are detailed in Table 3.

Table 6. Ranking of habitat water requirements.

Water Requirement Rank	Water Surface Area Percentage	Habitat Water Requirement (Billion m^3)
Maximum	80%	11.54
Suitable	50%	7.21
Minimum	30%	4.33

3.5. Calculation Results of Water Requirement for Groundwater Recharge

The coefficient of permeability of the Momoge Wetland is set to 0.005 m/d, the hydraulic gradient at 0.2, and the computational period at 180 days. The water requirement for groundwater recharge is equivalent to that of the habitat, with maximum, suitable, and minimum thresholds corresponding to water surface area percentages of 80%, 50%, and 30%, respectively. Precipitation between June and August accounts for approximately 68% of the annual total in Momoge Wetland, whereas runoff from May to September constitutes about 88%, marking this period as the critical water replenishment season. For this study, the replenishment period is established as 180 days. Based on the monthly runoff data collected at the Zhenlai station from 2000 to 2014, the allocated water requirement for groundwater recharge for the overwintering period (November to March of the following year), breeding period (April to June), and flood period (July to October) are 0.04, 0.21, and 0.75, respectively. The water requirement for groundwater recharge is calculated utilizing Formula (6), and the maximum, suitable, and minimum water requirements for the Momoge Wetland are specified in Table 7. The water requirements for groundwater recharge during different periods within the year are detailed in Table 3.

Table 7. Ranking of water requirements for groundwater recharge.

Water Requirement Rank	Water Surface Area Percentage	Water Requirement for Groundwater Recharge (Billion m^3)
Maximum	80%	2.08
Suitable	50%	1.30
Minimum	30%	0.78

4. Results

4.1. Annual Ecological Water Requirement of Momoge Wetland

Figure 2a illustrates the annual suitable ecological water requirement and threshold of the target level in the Momoge Wetland. To maintain the wetland's scale, promote the conservation of biodiversity, and ensure the stability of the ecosystem's functions and structure, the maximum annual ecological water requirement is calculated to be 24.14×10^8 m^3, the suitable requirement is 16.65×10^8 m^3, and the minimum requirement is 10.88×10^8 m^3. Among the various water requirement targets, promoting the conservation of biodiversity is the most critical for the Momoge Wetland, with a suitable ecological water requirement of 9.56×10^8 m^3, which accounts for 57% of the total suitable ecological water requirement. This is followed by maintaining the wetland's scale, which accounts for 43% of the total suitable ecological water requirement, while stability of the ecosystem's functions and structure contributes the least, accounting for only 3% of the total suitable ecological water requirement. The threshold for the ecological water requirement required for the conservation of biodiversity is the highest, ranging between 6.52×10^8 m^3 and $14.01 \times 10^8 m^3$.

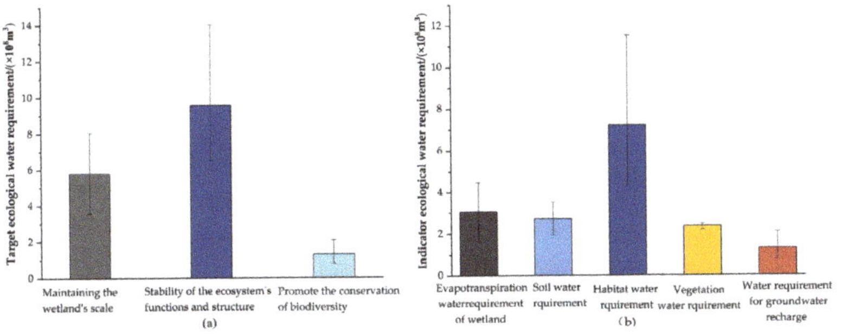

Figure 2. Annual suitable ecological water requirements and threshold of target (**a**) and indicator (**b**) level in the Momoge Wetland. Target ecological water requirements refer to maintaining the wetland's scale, promoting the conservation of biodiversity, and stability of the ecosystem's functions and structure. Indicators of ecological water requirements refer to evapotranspiration water requirement of wetland, soil water requirement, vegetation water requirement, habitat water requirement, and water requirement for groundwater recharge.

Figure 2b illustrates the annual suitable ecological water requirement and threshold of indicator level in the Momoge Wetland. The highest suitable ecological water requirement is for habitats, with an annual suitable ecological water requirement of 7.21×10^8 m^3, constituting 44% of the total suitable ecological water requirement. This is followed by the evapotranspiration water requirement of wetlands, which comprises 18% of the total water requirement. Soil water requirements, vegetation water requirements, and water requirements for groundwater recharge are distributed at 16%, 14%, and 8%, respectively. The ecological water thresholds for each index vary significantly, ranked from highest to lowest as follows: habitat water requirement (11.54×10^8 m^3–4.33×10^8 m^3), evapotranspiration water requirement of wetland (4.53×10^8 m^3–1.64×10^8 m^3), soil water requirement (3.52×10^8 m^3–1.94×10^8 m^3), water requirement for groundwater recharge (2.08×10^8 m^3–0.78×10^8 m^3), and vegetation water requirement (2.47×10^8 m^3–2.11×10^8 m^3).

4.2. Ecological Water Requirement of Momoge Wetland During Different Periods of the Year

Figure 3a illustrates the seasonal suitable ecological water requirement and threshold of the target in the Momoge Wetland. The ecological water requirement varies substantially across different periods, with the highest suitable ecological water requirement during the flood period (8.73×10^8 m^3), comprising 54% of the total annual suitable water requirement. This is followed by the breeding period (5.39×10^8 m^3), constituting 34% of the total suitable water requirement, while the overwintering period has a relatively lower suitable water requirement (1.92×10^8 m^3), constituting only 12% of the total annual suitable water requirement. In terms of water requirement targets, the differences in ecological water requirement across periods are not significantly distinct. The suitable ecological water requirements for maintaining the wetland's scale, promoting the conservation of biodiversity, and stability of the ecosystem's functions and structure are all higher during the flood period (3.35×10^8 m^3, 4.62×10^8 m^3, and 0.76×10^8 m^3, respectively) compared to the breeding period (1.93×10^8 m^3, 3.19×10^8 m^3, and 0.27×10^8 m^3, respectively) and the overwintering period (0.48×10^8 m^3, 1.39×10^8 m^3, and 0.05×10^8 m^3, respectively).

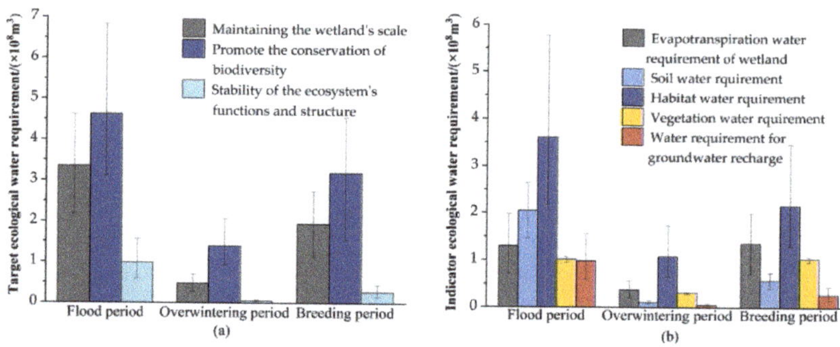

Figure 3. Seasonal suitable ecological water requirement and threshold of target (**a**) and indicator (**b**) in the Momoge Wetland. Target ecological water requirements refer to maintaining the wetland's scale, promoting the conservation of biodiversity, and the stability of the ecosystem's functions and structure. Indicators of ecological water requirements refer to evapotranspiration water requirements of wetland, soil water requirements, vegetation water requirements, habitat water requirements, and water requirements for groundwater recharge.

Figure 3b illustrates the seasonal suitable ecological water requirement and threshold of indicator in the Momoge Wetland. During the overwintering period, the predominant types of water requirement are habitat water requirement and evapotranspiration water requirement of wetland, with values of 1.08×10^8 m^3 and 0.37×10^8 m^3, respectively. In the breeding period, the suitable ecological water requirements for habitats, wetland evapotranspiration, and vegetation water increase rapidly compared to the overwintering period, with respective increases of 1.08×10^8 m^3, 0.99×10^8 m^3, and 0.72×10^8 m^3. In the flood period, while the evapotranspiration water requirements of wetland and vegetation water requirements decrease, the suitable ecological water requirements for other indices increase, with the largest increase seen in soil water requirement (1.48×10^8 m^3). From the perspective of water requirement thresholds, during the overwintering period, breeding period, and flood period, the habitat water requirement is the highest, at 1.73×10^8 m^3 to 0.65×10^8 m^3, 3.46×10^8 m^3 to 1.30×10^8 m^3, and 5.77×10^8 m^3 to 2.17×10^8 m^3, respectively.

4.3. The Status of Ecological Water Requirement Satisfaction in Momoge Wetland

We selected the annual runoff data for Momoge Wetland from 1979 to 2013, where the annual runoff in 1979 was 1.392 billion m^3, which was the lowest during this period and is referred to as the extreme dry year. The annual runoff in 1998 was 14.4 billion m^3, which was the highest during this period and is referred to as the extreme wet year.

Figure 4a illustrates the ratio of maximum ecological water requirements during the overwintering, breeding, and flood periods to the runoff of Momoge Wetland in 1979 and 1998. In the extreme dry year, the maximum ecological water requirement during the overwintering, breeding, and flood periods was substantially greater than the corresponding runoff in Momoge Wetland, being 3.14, 2.19, and 1.37 times the runoff, respectively. In the extreme wet year, while the maximum ecological water requirement during the flood period was only 10% of the runoff, the runoff during the overwintering and breeding periods could not meet the maximum ecological water requirement of the wetland.

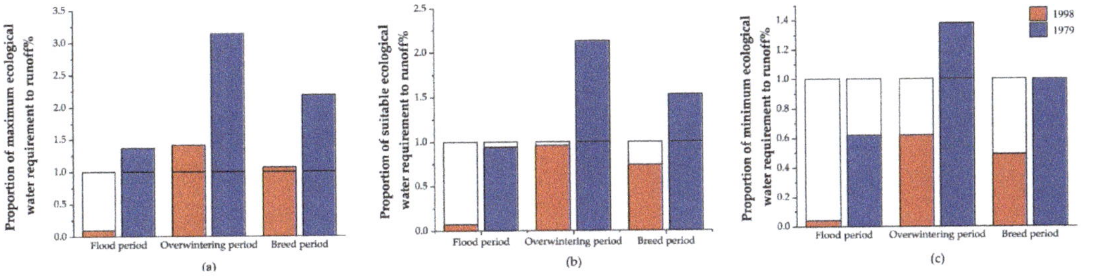

Figure 4. The maximum (**a**), suitable (**b**), and minimum (**c**) ecological water requirements for the wetland runoff seasons of Momoge Wetland in 1979 and 1998. The overwintering period, breeding period, and flood period refer to November to March of the following year, April to June, and July to October, respectively.

Figure 4b shows the proportion of suitable ecological water requirements during the overwintering, breeding, and flood periods relative to the Momoge Wetland's runoff in 1979 and 1998. In the extreme dry year, the suitable ecological water requirement during the overwintering and breeding periods was 2.13 and 1.53 times the runoff in Momoge Wetland, respectively. Although the suitable ecological water requirement during the flood period was lower than the runoff for that year, the proportion of the ecological base stream was 94%, which was far above normal levels. In the extreme wet year, the suitable ecological water requirement during the overwintering, breeding, and flood periods was lower than the runoff, but the proportions of the ecological base stream during the overwintering and breeding periods were 96% and 74%, respectively, which were higher than the normal levels. Thus, appropriate ecological water replenishment should be considered based on actual conditions.

Figure 4c shows the proportion of minimum ecological water requirement during the overwintering, breeding, and flood periods relative to the Momoge Wetland's runoff in 1979 and 1998. In the extreme dry year, the runoff in Momoge Wetland during the overwintering period could not meet the minimum requirement of the ecosystem. During the breeding period, the minimum ecological water requirement was equal to the runoff, while during the flood period, the proportion of the ecological base stream was 62%, which was also much higher than normal levels, suggesting the need for appropriate ecological water replenishment. In the extreme wet year, the runoff in Momoge Wetland could roughly meet the minimum ecological water requirement, with the proportions of the ecological base stream during the overwintering, breeding, and flood periods being 62%, 49%, and 4%, respectively.

4.4. Water Replenishment Plan for Momoge Wetland Based on Area Predictions

According to the existing literature [27], we utilized the grey prediction model to predict the area of the Momoge Wetland and assess the required water replenishment.

4.4.1. Extraction of Design Values for Abundant, Normal, and Dry Conditions of the Three Factors

Based on trend analysis, we calculate the correlation between precipitation in Momoge Wetland, Chaersen flow, Jiangqiao flow, annual average temperature, annual average evaporation, and the changes in the area of Momoge Wetland. Among these, the correlation coefficients with the highest correlation to the area changes of Momoge Wetland are rainfall, Jiangqiao flow, and Chalson flow, at 0.79, 0.67, and 0.62, respectively. Therefore, we recognized the precipitation in Momoge Wetland, the flow rate at Chaersen, and the flow rate at Jiangqiao as critical factors influencing the area of Momoge Wetland.

Four distinct water supplementation scenarios were established: (dry, dry, dry); (dry, dry, normal); (dry, normal, dry); and (normal, normal, normal). This study applies design frequency to classify the inflows, setting the threshold for distinguishing between wet and normal water conditions at the 25th percentile and between normal and dry conditions at the 75th percentile [22]. Based on historical data series of precipitation, Chaersen inflow, and Jiangqiao inflow, P-III frequency curves were generated, and the design values corresponding to abundant, normal, and dry conditions were calculated. The design values for precipitation, Chaersen inflow, and Jiangqiao inflow under abundant, normal, and dry conditions are displayed in Table 8.

Table 8. Design values corresponding to different frequencies.

Design Values	25%	50%	75%
Precipitation (mm)	398.5	345.2	310.6
Flow rate at Chaersen (billion m^3)	1.81	0.87	0.23
Flow rate at Jiangqiao (billion m^3)	225.8	180.6	138.4

4.4.2. Calculation of Ecological Water Replenishment Under Four Scenarios

Utilizing the grey prediction model, the design values corresponding to the four runoff scenarios were employed in Formulas (7)–(18) in Section 2.5 to predict the wetland area of the Momoge Wetland under different scenarios. The predicted values are shown in Table 8. Given that the wetland conditions are suboptimal in all four scenarios, water diversion is essential to fulfill the minimum ecological water requirement of the wetland [52]. Based on the predicted wetland area, the area of each wetland type was calculated proportionally, and the required water replenishment for each scenario was calculated using the formula in Section 2.4. The required water replenishment for each scenario is also displayed in Table 9.

Table 9. The predicted wetland area values of the four sceneries.

Design Value	Scenario One	Scenario Two	Scenario Three	Scenario Four
Wet and dry encounter combination	Dry-dry-dry	Dry-dry-normal	Dry-normal-dry	Normal-normal-normal
Predicted area (km^2)	1338.5	1340.9	1368.7	1387.7
Water replenishment (billion m^3)	0.70	0.49	0.68	0.36

4.4.3. Water Replenishment Recommendations

The Momoge Wetland primarily receives its water replenishment from the Nenjiang River, with a regular supply facilitated through the Nenjiang-to-Baicheng water diversion project. However, during periods of drought, this supply is insufficient to meet the wetland's water requirement. Therefore, it is recommended to establish additional water replenishment channels for the wetland, including the utilization of floodwater resources from the Tao'er River, the Yangshapao water storage project, and the water resources of the Chuoer River. These channels would expand the wetland protection area, improve the surrounding environment, and play a crucial role in restoring the Momoge Wetland. Furthermore, taking into account the operational requirements of reservoir managers, a multi-source water allocation strategy can be executed to optimize the fulfillment of the wetland's water requirement.

5. Discussion

This study, based on years of accumulated field data, calculated the overall ecological water requirements for the Momoge Wetland and emphasized the integrity of the wetland ecosystem. The calculation results can be effectively integrated with the ecological water regulation of Momoge Wetland, ensuring the ecological health of the wetland during the water resource regulation process. The implementation of wetland ecological water replenishment not only enhances the hydrological connectivity of the wetland, protects its hydrological factors and species diversity, but also maintains the structure and function of the wetland ecosystem [53]. Furthermore, it can promote regional economic development by increasing the connectivity between water bodies and expanding the capacity of wetland water resources, especially in the fields of tourism and leisure [54]. Wetland ecological water replenishment is also of great significance for climate regulation and water purification, all of which are key factors in achieving sustainable development [55]. Therefore, this study's proposal for estimating and forecasting ecological water requirements is practically valuable. It aids in the effective integration of the concepts, objectives, and processes related to the ecological water needs of wetland ecosystems.

Many scholars have conducted research on the ecological water requirement of the Momoge Wetland [26,56]. Due to the differences in wetland scale, calculation methods, and types of ecological water requirements considered by various scholars, the results have some uncertainties. In addition, the variations in runoff projections or the limitations of the grey prediction model can lead to some uncertainties. This study compares with representative studies on the estimation of ecological water requirements in Momoge Wetland, where the results of this study ($10.88{\sim}24.14 \times 10^8$ m^3) are relatively close to the results by Sun and Qu ($6.05{\sim}21.2 \times 10^8$ m^3) [26]. Both studies used the ecological function method for estimation, but there was a disagreement in the selection of wetland area. This study selects the ecological water requirements for the overall area of Momoge Wetland, while Sun's article calculates the minimum and suitable water requirements based on the core area and buffer zone of Momoge Wetland. The results differ significantly from those of Shen and Xu ($3.31{\sim}6.42 \times 10^8$ m^3) [56] because this study takes into account soil water requirements and biological habitat water requirements. After the study, it was found that Momoge Wetland is home to many rare animals and plants. Only by accurately understanding soil water requirements and biological habitat water requirements can we take effective protective measures and accurately assess the value of ecosystem services. Therefore, this study includes soil water requirements and biological habitat water requirements in the ecological water requirement index system. For calculating habitat water requirement, we applied the key species indicator method, selecting reed as the indicator species for the Momoge Wetland. The ecological function method used in this study calculates ecological water requirement from a macro perspective, while the key species indicator method links flora and fauna with the wetland from a micro perspective. By combining these two methods, we calculated an ecological water requirement that not only protects the ecosystem functions of the Momoge Wetland from a macro perspective but also takes into account the living environment of wetland species from a micro perspective.

For a long-term period, the Momoge Wetland has gradually degraded due to various factors, including natural environmental changes, population growth, and rapid economic development [34]. Ecological water replenishment is one of the most effective measures to address ecological problems [57]. Therefore, it is necessary to prioritize the full use of local water resources while also considering appropriate cross-basin water diversion projects and water transfer projects to protect the Momoge Wetland ecosystem [24]. The main available water sources for Momoge Wetland are the water introduced from the Nen River to the Ba River through the Nen-to-Ba Project and the runoff from the Tao'er River Basin. At present, the main water diversion project for the Momoge Wetland is the Nen-to-Ba Project. Although the current water supply from this project alleviates the wetland's water scarcity, it is still insufficient to meet the water requirement of the core zone during extreme dry years. Based on data analysis and the characteristics of floods in the Tao'er River

Basin [58], this study primarily investigates how to replenish the Momoge Wetland during small floods in the Tao'er River Basin. During such periods, we can reduce the speed of floodwaters by limiting reed harvesting within the wetland to enhance its flood detention capacity [30]. In addition, it is necessary to implement strict measures for converting farmland back to wetlands. Other measures, such as habitat enclosure, afforestation, and grazing bans, should be applied to control human-induced soil erosion, providing rare bird species with broader habitats and sufficient food sources. The health of wetland ecosystems is not only related to the amount of water but also related to hydrochemical elements. For example, the water quality in wetlands significantly affects the ecological patterns, processes, and hydrological service functions of the wetland ecosystem [59,60]. In Momoge Wetland, a large amount of farmland water carrying nutrients is discharged into the wetland, which is bound to affect the ecological security of the wetland [33]. Therefore, to carry out wetland ecological water replenishment in the future, we should not only consider the water requirement but also ensure the safety of water quality. These efforts will help maintain the ecological balance of the wetland and ensure the long-term stability of its water resources.

In this study, the calculation of ecological water requirement is based on the actual conditions of the Momoge Wetland, and the overall results are reliable. However, data limitations can have significant consequences for wetland water management. Incomplete or inaccurate data can lead to misinformed decisions regarding wetland water levels and flow regimes [61]. For instance, without detailed knowledge of historical water inflows and outflows, it becomes challenging to determine the optimal water levels for maintaining wetland biodiversity. This may result in the degradation of wetland habitats as water levels are either too high or too low. High water levels for an extended period can drown out native plant species and disrupt nesting sites for birds and other wildlife. Conversely, low water levels can lead to the drying up of wetland pools, reducing fish and amphibian habitats and increasing the vulnerability of the wetland to invasive species [62]. Data limitations also impede the ability to predict the impacts of external factors such as climate change and land use changes on wetland water systems [63]. Without accurate data on precipitation patterns, temperature trends, and upstream land development, it is difficult to model how these changes will affect wetland water availability and quality. This lack of predictive ability can prevent the timely implementation of adaptive management strategies to protect wetlands. Existing studies have shown that even with parameter uncertainty, reasonable model construction and scientific calculation methods can still yield results with practical reference values. For instance, Cui's [64] and Zhong's [65] articles have quantitatively calculated the ecological water requirement of wetlands using ecological function methods and ecological water level methods. The results indicate that, despite parameter uncertainty, the calculations can effectively guide ecological management and conservation practices. Meanwhile, Dong's [66] article emphasizes that model optimization and long-term data monitoring can not only reduce prediction errors but also enhance the reliability of the outcomes. The results of this study are consistent with the aforementioned research, further validating that parameter uncertainty does not affect the practical application value of the calculation results in wetland ecological water requirement studies, providing a scientific reference for future wetland ecological management and conservation.

When using the ecological function method to estimate ecological water requirements, the ecological function method provides a comprehensive framework for assessing the water requirements of different components in wetland ecosystems, including vegetation, soil, and habitats. This approach helps to identify and quantify the water needs of key ecological processes in wetland ecosystems, providing a scientific basis for wetland management and conservation. Although there may be issues of partial overlap in the calculation of wetland ecological water requirements, leading to potentially overestimated results, its flexibility and adaptability allow it to be adjusted according to the specific conditions of different wetlands. This means that by appropriate parameter adjustment and calibration, the impact of overlaps can be reduced, enhancing the accuracy of the calculation results. A substantial body of literature has validated the scientific feasibility of this method in wetland conservation practice. Future research can optimize models and apply integrated methods to reduce the impact of overlaps on the calculation results, thereby providing a more precise basis for wetland management.

Additionally, the conflict between water supply and requirement in the Momoge Wetland is prominent, making it difficult to coordinate water use interests [28]. When calculating ecological water requirements, it is necessary to consider a balance among economic, social, and environmental "win-win" outcomes, as well as uncertainties such as future ecological changes, to find a reasonable ecological water requirement that can be applied to practical water allocation schemes [67]. Furthermore, in the prediction of wetland ecological water replenishment, the grey prediction model provides a foundation, but the long-term correlation in the forecast may be limited by the incomplete integration of future climate change scenarios. To gain a more balanced perspective, it is essential to recognize the uncertainty inherent in the grey prediction model, particularly for long-term forecasting. Additionally, to enhance the accuracy of these predictions, we should consider integrating more climate variables and scenarios into our models [68]. Furthermore, strengthening our understanding of the climate system and exploring the mechanisms of how different climatic environments affect the ecological water needs of wetlands are important ways to address this limitation. Enhancing the precision of climate model simulations and forecasts further aids in overcoming these challenges [69]. Grey prediction models can enhance their ability to capture the dynamics of complex systems by introducing more variables and parameters. Especially when socio-economic factors are considered, the predictive power of the model can be significantly improved [70]. Moreover, grey prediction models inherently include the handling of uncertainty, making them still effective when dealing with systems that have incomplete information or are subject to change. Therefore, by comprehensively considering more variables, valuing socio-economic factors, and acknowledging the uncertainty in the forecasting process, the effectiveness and applicability of grey prediction models can be further enhanced.

6. Conclusions

In this study, we applied the ecological function method to assess the evapotranspiration water requirement of wetlands, soil water requirement, vegetation water requirement, habitat water requirement, and water requirement for groundwater recharge. Utilizing the three objective levels—maintaining the wetland's scale, promoting the conservation of biodiversity, and the stability of the ecosystem's functions and structure—we determined the ecological water requirement for various levels and temporal periods throughout the year. The key findings are as follows:

1. The maximum annual water ecological requirement of the Momoge Wetland is 24.14×10^8 m^3, the suitable ecological water requirement is 16.65×10^8 m^3, and the minimum ecological water requirement is 10.88×10^8 m^3.
2. The suitable ecological water requirement for the overwintering period, breeding period, and flood period is 1.92×10^8 m^3, 5.39×10^8 m^3, and 8.73×10^8 m^3, respectively. The suitable ecological water requirement for maintaining the wetland's scale, promoting the conservation of biodiversity, and stability of the ecosystem's functions and structure is 5.79×10^8 m^3, 9.56×10^8 m^3, and 1.3×10^8 m^3, respectively.

3. The required water replenishment for the Momoge Wetland under the four water encounter scenarios of dry-dry-dry, dry-dry-normal, dry-normal-dry, and normal-normal-normal are 0.70×10^8 m^3, 0.49×10^8 m^3, 0.68×10^8 m^3, and 0.36×10^8 m^3, respectively. In dry years, the existing water replenishment projects cannot fully meet the water storage requirement of the Momoge Wetland, highlighting the urgent need to adopt multi-source water replenishment methods to enhance replenishment efficiency.
4. Given the water scarcity in the study area, it is imperative to implement efficient water usage policies to reduce socio-economic water consumption while considering the spatial distribution of water resources in basin planning. Enhancing the efficiency of water resource utilization is crucial for ensuring the sustainability of water resources in the basin and has significant implications for biodiversity conservation, wetland restoration, and long-term water resource planning in China.

Author Contributions: H.M.: Writing—original draft, methodology, investigation, formal analysis, data curation. X.Z.: Writing—review and editing, conceptualization. Y.W.: Writing—review and editing, methodology, investigation, funding acquisition, formal analysis, data curation. X.P.: Methodology, investigation, funding acquisition, formal analysis, data curation. Z.L.: Writing—review and editing, project administration, conceptualization. Z.W.: Writing—review and editing, project administration, conceptualization. All authors have read and agreed to the published version of the manuscript.

Funding: This work was supported by The National Key Research and Development Program of China (2021YFC3200203), and the National Natural Science Foundation of China (421010514, U2243230, 422070881, and 42371169).

Data Availability Statement: Data will be made available on request.

Conflicts of Interest: The authors declare that they have no known competing financial interests or personal relationships that could have appeared to influence the work reported in this paper.

References

1. Salimi, S.; Almuktar, S.A.A.A.; Scholz, M. Impact of climate change on wetland ecosystems: A critical review of experimental wetlands. *J. Environ. Manag.* **2021**, *286*, 112160. [CrossRef] [PubMed]
2. Acharya, A.; Sharma, M.L.; Bishwakarma, K.; Dahal, P.; Chaudhari, S.K.; Adhikari, B.; Neupane, S.; Pokhrel, B.N.; Pant, R.R. Chemical Characteristics of the Karmanasha River Water and Its Appropriateness for Irrigational Usage. *J. Nepal Chem. Soc.* **2020**, *41*, 94–102. [CrossRef]
3. Fluet-Chouinard, E.; Stocker, B.D.; Zhang, Z.; Malhotra, A.; Melton, J.R.; Poulter, B.; Kaplan, J.O.; Goldewijk, K.K.; Siebert, S.; Minayeva, T.; et al. Extensive global wetland loss over the past three centuries. *Nature* **2023**, *614*, 281. [CrossRef] [PubMed]
4. Davidson, N.C. How much wetland has the world lost? Long-term and recent trends in global wetland area. *Mar. Freshwater Res.* **2014**, *65*, 934. [CrossRef]
5. Mitsch, W.J.; Gosselink, J.G. *Wetlands*; John Wiley and Sons, Inc.: Hoboken, NJ, USA, 2015; ISBN 9781118676820.
6. Thapa, B.; Pant, R.R.; Thakuri, S.; Pond, G. Assessment of spring water quality in Jhimruk River Watershed, Lesser Himalaya, Nepal. *Environ. Earth Sci.* **2020**, *79*, 1–14. [CrossRef]
7. Tabari, H. Climate change impact on flood and extreme precipitation increases with water availability. *Sci. Rep.* **2020**, *10*, 13768. [CrossRef]
8. Wang, X.; Chen, Y.; Li, Z.; Fang, G.; Wang, F.; Hao, H. Water resources management and dynamic changes in water politics in the transboundary river basins of Central Asia. *Hydrol. Earth Syst. Sci.* **2021**, *25*, 3281–3299. [CrossRef]
9. Franco-Torres, M.; Rogers, B.C.; Harder, R. Articulating the new urban water paradigm. *Crit. Rev. Environ. Sci. Technol.* **2021**, *51*, 2777–2823. [CrossRef]
10. Xu, Y.; Wang, Y.; Li, S.; Huang, G.; Dai, C. Stochastic optimization model for water allocation on a watershed scale considering wetland's ecological water requirement. *Ecol. Indic.* **2018**, *92*, 330–341. [CrossRef]
11. Zhao, F.; Pang, A.P.; Li, C.H. Progress in ecological water requirements in the mainstream and estuary ecosystem of the Yellow River Basin. *Acta Ecol. Sin.* **2021**, *41*, 6289–6301.
12. Hao, X.; Zhao, Z.; Fan, X.; Zhang, J.; Zhang, S. Evaluation method of ecological water demand threshold of natural vegetation in arid-region inland river basin based on satellite data. *Ecol. Indic.* **2023**, *146*, 109811. [CrossRef]

13. Tharme, R.E. A global perspective on environmental flow assessment: Emerging trends in the development and application of environmental flow methodologies for rivers. *River Res. Appl.* **2003**, *19*, 397–441. [CrossRef]
14. Liu, Z.; Zhang, Z.; Ma, H.; Gao, J.; Su, Z.; Wang, H.; Li, G.; Ding, X. Optimal scheduling study of cascade hydropower stations to ensure ecological flow requirements-A case study of the Gorge Section of the Yongding River, China. *Sci. Total Environ.* **2024**, *953*, 175977. [CrossRef]
15. Cui, B.; Tang, N.; Zhao, X.; Bai, J. A management-oriented valuation method to determine ecological water requirement for wetlands in the Yellow River Delta of China. *J. Nat. Conserv.* **2009**, *17*, 129–141. [CrossRef]
16. Cui, B.; Yang, Z.F. Water consumption for eco-environmental aspect on wetlands. *Acta Sci. Circumstantiae* **2002**, *22*, 219–224. [CrossRef]
17. Wu, H.; Shi, P.; Qu, S.; Zhang, H.; Ye, T. Establishment of watershed ecological water requirements framework: A case study of the Lower Yellow River, China. *Sci. Total Environ.* **2022**, *820*, 153205. [CrossRef] [PubMed]
18. Liu, J.Y.; Mei, Z.H.; Gai, H.Y.; Zhang, Y.X.; Mei, R.X. Ecological water replenishment calculation and water replenishment scheme analysis in Yellow River Delta. *Water Resour. Hydropower Eng.* **2024**, *55*, 222–227. [CrossRef]
19. Mao, B.; Wang, X.; Liao, Z.; Miao, Y.; Yan, S. Spatiotemporal variations and tradeoff-synergy relations of ecosystem services under ecological water replenishment in Baiyangdian Lake, North China. *J. Environ. Manag.* **2023**, *343*, 118229. [CrossRef] [PubMed]
20. Huang, F.; Chunyu, X.; Zhang, D.; Chen, X.; Ochoa, C.G. A framework to assess the impact of ecological water conveyance on groundwater-dependent terrestrial ecosystems in arid inland river basins. *Sci. Total Environ.* **2020**, *709*, 136155. [CrossRef] [PubMed]
21. Wu, Y.; Xu, J.Q.; Qiu, W.S.; Wang, M.Y.; Li, S.; Shi, Y.; Wei, J.H. Study on ecological water replenishment scheduling scheme of Chaobai River in Beijing section. *Water Resour. Hydropower Eng.* **2023**, *54*, 180–189. [CrossRef]
22. Jiang, W.; Zhuang, W.; Wen, C.; Pang, Y.; Chao, J. Calculation of ecological water replenishment in small coastal watersheds based on environmental capacity constraints: A case study of eastern Xiamen, China. *Ecol. Indic.* **2023**, *155*, 111034. [CrossRef]
23. Zhang, X.; Duan, B.; He, S.; Lu, Y. Simulation study on the impact of ecological water replenishment on reservoir water environment based on Mike21--Taking Baiguishan reservoir as an example. *Ecol. Indic.* **2022**, *138*, 108802. [CrossRef]
24. Jiang, H.; Wen, Y.; Zou, L.; Wang, Z.; He, C.; Zou, C. The effects of a wetland restoration project on the Siberian crane (*Grus leucogeranus*) population and stopover habitat in Momoge National Nature Reserve, China. *Ecol. Eng.* **2016**, *96*, 170–177. [CrossRef]
25. Ye, Z.; Yang, Y.; Zhou, H.; Guo, B. Ecological Water Rights of the Bosten Lake Wetlands in Xinjiang, China. *Wetlands* **2020**, *40*, 2597–2607. [CrossRef]
26. Sun, X.M.; Qu, Y.G. Study on Ecological Water Requirement of Momoge Wetland Ecological Environment. *Water Resour. Hydropower Northeast China* **2007**, *25*, 54–55.
27. Shi, W.J. *Study on the Changes of Spatial Pattern and the Regulations Ecological Water Demand in Momoge Wetland*; Dalian University of Technology: Dalian, China, 2016.
28. Han, J.; Wang, D.; Zhang, S. Momoge Internationally Important Wetland: Ecosystem Integrity Remote Assessment and Spatial Pattern Optimization Study. *Land* **2022**, *11*, 1344. [CrossRef]
29. Wen, D.; Su, L.; Hu, Y.; Xiong, Z.; Liu, M.; Long, Y. Surveys of Large Waterfowl and Their Habitats Using an Unmanned Aerial Vehicle: A Case Study on the Siberian Crane. *Drones* **2021**, *5*, 102. [CrossRef]
30. Zhang, M.; Zhang, D.; Qi, Q.; Tong, S.; Wang, X.; An, Y.; Lu, X. Flooding effects on population and growth characteristics of Bolboschoenus planiculmis in Momoge wetland, Northeast China. *Ecol. Indic.* **2022**, *137*, 108730. [CrossRef]
31. Zhao, Y.; Mao, D.; Zhang, D.; Wang, Z.; Du, B.; Yan, H.; Qiu, Z.; Feng, K.; Wang, J.; Jia, M. Mapping Phragmites australis Aboveground Biomass in the Momoge Wetland Ramsar Site Based on Sentinel-1/2 Images. *Remote Sens.* **2022**, *14*, 694. [CrossRef]
32. Liu, Y. Mechanism and Pattern Analysis of Wetland Hydrological Connectivity: Taking Momoge National Nature as an Example. Ph.D. Thesis, Jilin University, Jilin, China, 2020.
33. Gao, J.; Deng, G.; Jiang, H.; Ma, Q.; Wen, Y.; He, C.; Guo, Y.; Cao, Y. Water quality management of micro swamp wetland based on the "source-transfer-sink" theory: A case study of Momoge Swamp Wetland in Songnen Plain, China. *J. Clean. Prod.* **2024**, *446*, 141450. [CrossRef]
34. Zhang, P.; Liao, B. The Dynamic Changes of Landscape Connectivity in Momoge Wetland based on Long Time Series Remote Sensing Data. In Proceedings of the IGARSS 2024—2024 IEEE International Geoscience and Remote Sensing Symposium, Athens, Greece, 7–12 July 2024.
35. Cui, G.; Liu, Y.; Tong, S. Analysis of the causes of wetland landscape patterns and hydrological connectivity changes in Momoge National Nature Reserve based on the Google Earth Engine Platform. *Arab. J. Geosci.* **2021**, *14*, 1–16. [CrossRef]
36. Zhang, X.; Liu, L.; Zhao, T.; Wang, J.; Liu, W.; Chen, X. Global annual wetland dataset at 30 m with a fine classification system from 2000 to 2022. *Sci. Data* **2024**, *11*, 1–12. [CrossRef] [PubMed]
37. Yan, F.; Wei, S.; Zhang, J.; Hu, B. Depth-to-bedrock map of China at a spatial resolution of 100 meters. *Sci. Data* **2020**, *7*, 1–13. [CrossRef] [PubMed]

38. Zheng, C.; Jia, L.; Hu, G. Global Land Surface Evapotranspiration Monitoring by ETMonitor Model Driven by Multi-source Satellite Earth Observations. *J. Hydrol.* **2022**, *613*, 128444. [CrossRef]
39. Ministry of Water Resources, P.R. China. China Water Statistical Yearbook Editorial Board. In *China Water Statistical Yearbook 2023*; China Water & Power Press: Beijing, China, 2023.
40. *GB/T 24708-2009*; Standardization Administration of the People's Republic of China. Wetland Classification. Standards Press of China: Beijing, China, 2009.
41. Cui, B.S.; Li, Y.H.; Yang, Z.F. Management-oriented ecological water requirement for wetlands in the Yellow River Delta. *Acta Ecol. Sin.* **2005**, *25*, 606–614.
42. Qiu, M.Q.; Han, M.; Jiao, C.T.; Song, S.; Liu, Y. Estimation of Ecological Water Requirement in the Yellow River Delta Wetland. *Acta Ecol. Sin.* **2023**, *43*, 9096–9105. [CrossRef]
43. Li, L.J.; Li, J.Y.; Liang, L.Q.; Liu, Y. Method for calculating ecological water storage and ecological water requirement of marsh. *J. Geogr. Sci.* **2009**, *19*, 427–436. [CrossRef]
44. Liu, C.M.; Wang, H.X. *The Interface Processes of Water Movement in the Soil-Crop-Atmosphere System and Water-Saving Regulation*; Science Press: Beijing, Chian, 1999; ISBN 7-03-007469-6.
45. Bai, T.; Liu, D.; Deng, M. Multi-scale ecological operation model of reservoir group coupled with ecological infiltration irrigation. *Agr. Water Manag.* **2022**, *270*, 107723. [CrossRef]
46. Wang, L.; Yang, X. Estimation of Environmental Water Requirements via an Ecological Approach: A Case Study of Yongnian Wetland, Haihe Basin, China. In *Sustainable Development of Water Resources and Hydraulic Engineering in China*; Springer: Cham, Switzerland, 2019; pp. 377–386. [CrossRef]
47. Yu, X.; Zhu, W.; Wei, J.; Jia, S.; Wang, A.; Huang, Y.; Zhao, Y. Estimation of ecological water supplement for typical bird protection in the Yellow River Delta wetland. *Ecol. Indic.* **2021**, *127*, 107783. [CrossRef]
48. Liu, S.; Lin, Y. *Grey Systems*; Springer: Berlin/Heidelberg, Germany, 2010; ISBN 9783642161575.
49. You, K.; Kang, N.; Fu, J.; Gao, Y.; Qian, W.; Wang, J. Research on Forecast of Water Demand in Jinzhou Based on Grey Model. *IOP Conf. Ser. Earth Environ. Sci.* **2021**, *770*, 12037. [CrossRef]
50. Gao, H.K.; Tao, Y.Z.; Yang, J. Prediction of domestic water consumption based on improved GM(1,1) model. *J. Hefei Univ. Technol.* **2024**, *47*, 387–391. [CrossRef]
51. Lindenmayer, D.B.; Bowd, E.; Youngentob, K.; Evans, M.J. Identifying biodiversity surrogates and management indicator species for tall, WET Forests: A case study of australian arboreal marsupials. *Ecol. Indic.* **2024**, *166*, 112297. [CrossRef]
52. Gao, J.; Gao, Y.; Zhao, G.; Hörmann, G. Minimum ecological water depth of a typical stream in Taihu Lake Basin, China. *Quatern Int.* **2010**, *226*, 136–142. [CrossRef]
53. Gupta, R.; Chembolu, V.; Marjoribanks, T.I.; Dutta, S. Assessing the efficacy of hydro-ecological based wetland management approach for flood resilience of a large river catchment. *J. Hydrol.* **2024**, *641*, 131761. [CrossRef]
54. Zhou, J.; Hou, Q.; Li, W. Spatial resilience assessment and optimization of small watershed based on complex network theory. *Ecol. Indic.* **2022**, *145*, 109730. [CrossRef]
55. Feng, H.; Liu, M.; Xu, M.; Zhang, M.; Mo, L.; Chen, T.; Tan, X.; Liu, Z. Study on the integrated protection strategy of water environment protection: The case of Hainan Province of China. *Environ. Technol. Inno* **2021**, *24*, 101990. [CrossRef]
56. Shen, H.A.; Xu, M.X.; Wang, F.Z. Calculation of wetland ecological water demand in Momoge Nature reserve. *Water Resour. Hydropower Northeast China* **2009**, *27*, 44–47.
57. Xu, H.; Zhang, D.; Wu, Y.; Qi, P.; Wang, X. Suitable Ecological Water Demand for Wetlands Restored to Different Historical Periods in a Latitude area and their Response to Changing Environments. *Water Resour. Manag.* **2024**, *38*, 5683–5700. [CrossRef]
58. Lang, H.L. *Water Resource Management and Protection in Momoge Wetland*; Jilin University: Jilin, China, 2019.
59. Cui, S.; Yu, T.; Zhang, F.; Fu, Q.; Hough, R.; An, L.; Gao, S.; Zhang, Z.; Hu, P.; Zhu, Q.; et al. Understanding the risks from diffuse pollution on wetland eco-systems: The effectiveness of water quality classification schemes. *Ecol. Eng.* **2020**, *155*, 105929. [CrossRef]
60. Xu, J.; Bai, Y.; You, H.; Wang, X.; Ma, Z.; Zhang, H. Water quality assessment and the influence of landscape metrics at multiple scales in Poyang Lake basin. *Ecol. Indic.* **2022**, *141*, 109096. [CrossRef]
61. Allen, R.G.; Pereira, L.S.; Howell, T.A.; Jensen, M.E. Evapotranspiration information reporting: I. Factors governing measurement accuracy. *Agr. Water Manag.* **2011**, *98*, 899–920. [CrossRef]
62. Liu, X.; Yang, W.; Fu, X.; Li, X. Determination of the ecological water levels in shallow lakes based on regime shifts: A case study of China's Baiyangdian Lake. *Ecohydrol. Hydrobiol.* **2024**, *24*, 931–943. [CrossRef]
63. Erwin, K.L. Wetlands and global climate change: The role of wetland restoration in a changing world. *Wetl. Ecol. Manag.* **2009**, *17*, 71–84. [CrossRef]
64. Cui, L.J.; Bao, D.M.; Xiao, H. Calculation Methods of Wetland Ecological Water and Case Study. *J. Soil Water Conserv.* **2005**, *29*, 147–151. [CrossRef]

65. Zhong, P.; Yang, Z.F.; Cui, B.S.; Liu, J.L. Studies on water resource requirement for eco-environmental use of the Baiyangdian Wetland. *Acta Sci. Circumstantiae* **2005**, *25*, 1119–1126. [CrossRef]
66. McLaughlin, D.L.; Cohen, M.J. Realizing Ecosystem Services: Wetland Hydrologic Function along a Gradient of Ecosystem Condition. *Ecol. Appl.* **2013**, *23*, 1619–1631. [CrossRef]
67. Huang, J.; Yang, H.; He, W.; Li, Y. Ecological Service Value Tradeoffs: An Ecological Water Replenishment Model for the Jilin Momoge National Nature Reserve, China. *Int. J. Environ. Res. Public Health* **2022**, *19*, 3263. [CrossRef] [PubMed]
68. Tian, Z.W.; Qian, R.L. Chinese Water Demand Forecast Based on iTransformer Model. *IEEE Access* **2024**, *12*, 115853–115867. [CrossRef]
69. Ling, Z.; Shu, L.; Wang, D.; Lu, C.; Liu, B. Assessment and projection of the groundwater drought vulnerability under different climate scenarios and land use changes in the Sanjiang Plain, China. *J. Hydrol. Reg. Stud.* **2023**, *49*, 101498. [CrossRef]
70. Zhang, X.; Dang, Y.; Ding, S.; Wang, J. A novel discrete multivariable grey model with spatial proximity effects for economic output forecast. *Appl. Math. Model.* **2023**, *115*, 431–452. [CrossRef]

water

MDPI

Article

Awareness and Behaviors of Beijing Residents Regarding Wetland Conservation

Lichun Mo and Botao Yan *

School of Economics and Management, North China Institute of Science and Technology, Beijing 065201, China; lichunmo@ncist.edu.cn
* Correspondence: yanbotao@ncist.edu.cn

Abstract: Wetlands are among the most important ecosystems worldwide, playing an irreplaceable role in maintaining ecological balance and ensuring human well-being. This study conducted a questionnaire survey of 1008 residents across 16 districts in Beijing to systematically analyze their awareness, attitudes, and behaviors regarding wetland conservation. The results indicate that the majority of residents hold a positive attitude towards wetland conservation, with over 90% supporting the expansion of wetland areas and 82% considering wetland protection more important than economic development. Regarding willingness to pay, residents tend to prefer small donations, with the highest proportion willing to contribute less than CNY 100 annually, while 92.5% expressed a willingness to support wetland conservation through volunteer work. Residents' behaviors are significantly influenced by the geographical location and functional configuration of wetlands, with higher visit rates observed at wetlands closer to urban areas, such as Shichahai Wetland and Summer Palace Wetland. This study also reveals that public awareness of wetland ecological functions is lacking, particularly in terms of biodiversity conservation and water quality improvement. The findings provide a scientific basis for enhancing wetland conservation policies and improving public awareness and actions for wetland protection in Beijing.

Keywords: wetland conservation; public awareness; willingness to pay; urban wetlands; environmental behavior

1. Introduction

Wetlands are widely recognized by international ecologists as one of the most important ecosystems globally, playing an irreplaceable role in maintaining ecological balance and ensuring human well-being [1]. Urban wetlands are considered key natural units for maintaining urban sustainability while also being highly vulnerable ecological units [2]. Urban wetlands not only provide diverse habitats for wildlife but also offer a range of ecosystem services, including climate regulation, air quality improvement, water quality improvement, flood regulation, and opportunities for recreation and cultural education [3]. However, with the rapid expansion of cities, the reduction in urban wetland areas and the degradation of their functions have become pressing issues that require urgent solutions. Currently, research on wetlands tends to focus more on coastal, lake, or river wetlands, covering topics such as the spatiotemporal distribution of these wetland types, ecosystem assessment, ecological restoration, carbon cycling, and management systems. In comparison, research on urban wetland conservation, particularly in megacities like Beijing, is still limited, and this study aims to fill this gap by providing a detailed examination of public awareness and participation in urban wetland conservation in Beijing. The novelty of this case study lies in its specific focus on addressing the unique pressures and challenges faced by urban wetlands in a rapidly urbanizing megacity while also considering the ecological characteristics of these wetlands that affect public perception. Research on urban wetland conservation remains relatively scarce, insufficient to meet the real-world needs of ecological protection and sustainable development in the context of urbanization.

In Beijing, the total wetland area is 62,100 hectares, accounting for approximately 3.79% of the city's total area [4]. Due to economic development, accelerated urbanization, and weak awareness of wetland protection, Beijing has seen a reduction in wetland areas of approximately 194,700 hectares since the 1950s, with large areas of wetlands shrinking or even disappearing [5]. In recent years, the Beijing government has undertaken substantial efforts to protect wetland resources by implementing various conservation and restoration measures, such as establishing wetland nature reserves, wetland parks, and micro wetlands. These efforts reflect the city's commitment to protecting and restoring wetlands and their functions, promoting the sustainable utilization of ecological resources. However, the effective protection and management of urban wetlands depend not only on policy implementations but also on active public awareness and perceptions [6]. Public perception of the wetland environment is the result of complex interactions between humans and the environment, involving emotional and cognitive processes linked to values, cultural backgrounds, attitudes, and preferences [7]. A study based on the Personal Construct Theory explored public perceptions of freshwater wetlands in Victoria, Australia, and found that public perceptions encompassed not only the landscape patterns of wetlands but also their biology, habitats, ecosystems, and ecological processes [8]. Therefore, the public's awareness and attitudes toward urban wetlands can provide strong support for the formulation and effective implementation of wetland protection policies, as well as the development of scientific and sustainable wetland management models [9]. A study on urban wetlands in Mexico assessed residents' perceptions of wetland characteristics, revealing that in addition to socio-economic and environmental variables, regional management variables—such as canals, urban housing, and irregular settlements—significantly influenced residents' perceptions of urban wetlands [10]. Moreover, research on Xuan Thuy National Park in Vietnam found that surrounding community residents generally held positive attitudes toward wetland conservation but had low awareness of the potential threats wetlands faced and the park's regulations. Age, length of residence, and education level were identified as major factors influencing conservation attitudes [11]. However, extensive research has been conducted on coastal and rural wetlands, but the unique pressures and public attitudes surrounding urban wetlands in megacities like Beijing remain understudied.

Public awareness and behavior are key to maintaining the health of urban wetland ecosystems. Understanding the willingness of 377 residents to participate in urban wetland conservation and the factors that influence this willingness is of great significance [12]. Based on this, the present study conducted a questionnaire survey among 1008 residents in 16 districts of Beijing, providing an in-depth analysis of the awareness and behavioral tendencies of Beijing residents regarding urban wetland conservation. The objectives of this study are to (1) assess Beijing residents' attitudes and behavioral tendencies toward wetland conservation; (2) identify the key demographic and socio-economic factors influencing public awareness and willingness to engage in conservation efforts; and (3) explore effective strategies for enhancing public participation in urban wetland conservation. The findings of this case study will provide insights for the innovation of urban wetland management models and the design of long-term wetland conservation mechanisms in China.

2. Materials and Methods

2.1. Research Area

Beijing is located in the northwest corner of the North China Plain and serves as the political and cultural center of the country. It is also an international city with a rich historical heritage and modern characteristics. Apart from its southeastern border with Tianjin, Beijing is surrounded by Hebei Province, and it is approximately 150 km southeast of the Bohai Sea. The city's geographic coordinates range from 39°26′ to 41°04′ N latitude and 115°25′ to 117°30′ E longitude. Beijing covers a total land area of 16.4 million hectares, of which mountainous areas account for 10.1 million hectares, or 61.6% of the total area, while plains cover 6.3 million hectares, representing 38.4% of the total area. The city's terrain is diverse, primarily composed of hills, plains, river valleys, and basins. Beijing experiences a temperate semi-humid continental monsoon climate, characterized by most rainfall occurring in summer (June to August) and autumn (September to November), with winters (December to February) being predominantly dry and cold. According to data from the National Bureau of Statistics, Beijing's urbanization rate was 29.5% in 1984, but by 2020, it had risen to approximately 87.5%, indicating significant urban development over the past few decades. Additionally, Beijing boasts numerous cultural heritage sites and tourism resources, attracting millions of visitors each year, contributing significantly to the city's tourism industry and cultural economy. As of the 2022 survey, the total area of wetlands larger than 400 square meters in Beijing was 62,100 hectares, accounting for 3.79% of the city's total area. Beijing's wetlands are mainly distributed across the five major water systems (Figure 1), including the Yongding River, Chaobai River, and Beiyun River, as well as over 20 large- and medium-sized reservoirs. Wetlands are found in all 16 districts of Beijing, with significant concentrations in Miyun District, Tongzhou District, and Fangshan District. The largest drinking water source in Beijing, the Miyun Reservoir, is also located in a wetland area. Wetland nature reserves are distributed in Yanqing District, Shunyi District, Fangshan District, and Huairou District.

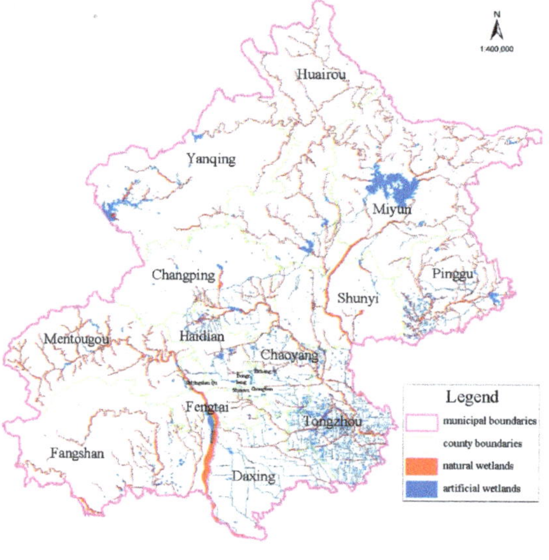

Figure 1. Distribution of wetlands in Beijing.

2.2. Sample Selection

The selected regions were chosen based on key criteria: the diversity of wetland types (including reserves, parks, and micro-wetlands), ensuring a broad representation of ecosystems; higher wetland concentrations for geographical diversity; ecological importance, with a focus on Miyun, Tongzhou, and Fangshan; and urbanization levels, allowing for a comparison of public awareness across different urban settings. This study focused on the 16 districts of Beijing, where we conducted on-site investigations into various types of wetlands within these districts. After designing the questionnaire, a small-scale pilot survey was conducted to refine and improve the questionnaire, followed by extensive on-site surveys in different types of wetlands across the 16 districts. A total of 97 locations were randomly selected across the districts for the survey, and 1008 valid questionnaires were collected. Of these, 420 were collected through on-site surveys, while 588 were collected via online questionnaires (Table 1). The on-site surveys were conducted in public spaces near urban wetlands, while the online surveys were distributed to a broader audience through social media and local community networks. The questionnaire responses were finalized by October 2022.

Table 1. Sample collection information.

District	Wetlands Surveyed	Residents Surveyed
Xicheng	6	167
Chaoyang	9	137
Fengtai	6	45
Shijingshan	2	21
Haidian	16	213
Shunyi	2	55
Tongzhou	9	41
Daxing	7	84
Fangshan	8	32
Dongcheng	1	3
Mentougou	10	60
Changping	5	29
Pinggu	3	24
Miyun	3	34
Huairou	6	26
Yanqing	4	37

The number of residents surveyed in each wetland was not consistent, as the selection was based on the distribution of wetlands. Selection criteria for participants included demographic characteristics (age, gender, education, occupation) to reflect Beijing's socio-economic diversity, length of residence to gauge familiarity with local wetlands, and frequency of interaction with wetlands (e.g., visits to parks and reserves) to assess exposure's effect on awareness. These criteria align with international studies on wetland and environmental awareness [8,13].

Wetlands in areas with a higher number of wetland sites were selected in greater quantity for surveys, while areas with fewer wetland sites had fewer survey locations. To address the uneven distribution of wetlands across districts, a stratified random sampling approach was employed to ensure proportional representation from districts with higher concentrations of wetlands. This approach aimed to ensure that wetlands with higher ecological importance or greater public exposure were well-represented in the sample. For example, Dongcheng and Shijingshan, which have fewer wetland sites, naturally had a smaller number of residents surveyed compared to districts with more abundant wetland resources.

2.3. Data Analysis

This study applied statistical methods to analyze the relationships between demographic characteristics, residents' awareness, attitudes, and behaviors toward urban wetland conservation. Descriptive statistics were used to summarize respondents' basic characteristics, including age, education level, income, and responses to key questions. Multivariate Analysis of Variance (MANOVA) was used to examine the relationships between individual socio-economic characteristics and residents' awareness, attitudes, and behaviors toward wetland conservation. Principal Component Analysis (PCA) was applied to identify predictors of residents' willingness to pay for wetland conservation and their willingness to volunteer, with independent variables including age, income, education.

3. Results and Analysis

3.1. Basic Characteristics of Residents and Their Households

This study first described the basic characteristics of the residents and their households (Table 2), including participants' gender, age, educational background, marital status, health status, occupation, Beijing household registration status, district of residence, homeownership in Beijing, length of residence in Beijing, and monthly post-tax income.

Table 2. Basic characteristics of the sampled residents and their families.

Indicator	Category	Proportion (%)	Indicator	Explanation	Proportion (%)
Gender	Male	40.77	Residential Area	Xicheng	6.85
	Female	59.23		Chaoyang	15.67
Age	Under 19	0.69		Fengtai	7.04
	19–35	70.54		Shijingshan	5.75
	36–50	21.03		Haidian	25.89
	51–69	7.74		Shunyi	4.76
Education Level	Vocational School	2.78		Tongzhou	4.66
	Associate Degree	10.02		Daxing	6.15
	Bachelor's Degree	61.01		Fangshan	4.96
	Master's Degree	17.36		Dongcheng	1.79
Marital Status	Married	51.59		Mentougou	4.46
	Unmarried	47.82		Changping	5.65
Health Condition	Very Poor	0.10		Pinggu	1.49
	Poor	3.77		Miyun	2.78
	Average	17.86		Huairou	2.08
	Good	58.63		Yanqing	1.88
	Very Good	19.64	Home Ownership in Beijing	No Property	47.92
Occupation	Farmer	0.60		Owns one property	44.74
	Laborer	6.15		Owns two properties	6.55
	Company Employee	50.20		Owns three or more properties	0.79
	Public Institution Employee	8.63	Years Living in Beijing	Less than 1 year	2.88
	Civil Servant	3.47		1 to 5 years	24.80
	Self-Employed	3.08		5 to 10 years	21.23
	Freelancer	2.48		10 to 15 years	8.73
	Student	20.14		More than 15 years	42.36
	Retired	3.47	Monthly Post-Tax Income	CNY 4000/month or below	19.05
	Active Military	0.10		CNY 4001 ~ 8000/month	18.85
	Other	1.49		CNY 8001 ~ 12,000/month	23.12
	Unemployed	0.20		CNY 12,001 ~ 16,000/month	21.53
Beijing Household Registration	Yes	56.35		CNY 16,001 ~ 20,000/month	9.52
	No	43.75		CNY 20,001 ~ 24,000/month	3.57
				CNY 24,001 ~ 28,000/month	1.59
				CNY 28,001 ~ 32,000/month	0.79
				More than CNY 32,000/month	1.98

In the survey sample, the proportion of male residents was slightly lower than that of female residents. Most respondents were between 19 and 35 years old, with the majority holding a bachelor's degree. The proportions of married and unmarried residents were roughly equal. Nearly 80% of respondents considered their health to be good or very good. In terms of employment, most participants were company employees, followed by students. The proportion of residents with a Beijing hukou was slightly higher than those without, aligning with the fact that more than half of the respondents owned homes in Beijing. Over half of the residents had lived in Beijing for more than 10 years. Most residents resided in Haidian, Chaoyang, Fengtai, Xicheng, and Daxing districts. Monthly post-tax income was concentrated in four ranges: CNY 12,000–16,000, CNY 8001–12,000, CNY 4001–8000, and CNY 4000 or below, reflecting the current socio-economic reality of Beijing.

3.2. Residents' Awareness and Attitudes Toward Current Wetland Conservation

Residents' perceptions of wetland conservation were assessed based on indicators such as air quality, biodiversity richness, water quality, landscaping conditions, urban cleanliness, and wetland protection status (Table 3).

Table 3. Sample residents' attitudes toward wetland conservation status.

Indicator	Category	Proportion (%)	Indicator	Explanation	Proportion (%)
Air Quality	Very poor	0.60	**Biodiversity Richness**	Very poor	0.69
	Poor	6.45		Poor	6.45
	Average	21.13		Average	47.08
	Good	49.80		Good	20.87
	Very good	22.02		Very good	24.90
Water Quality	Very poor	1.09	**Landscaping Conditions**	Very poor	0.50
	Poor	5.26		Poor	2.38
	Average	25.69		Average	14.88
	Good	37.90		Good	48.21
	Very good	30.06		Very good	34.03
Urban Cleanliness	Very poor	0.60	**Wetland Protection Status**	Very poor	0.60
	Poor	2.48		Poor	3.17
	Average	18.55		Average	19.05
	Good	50.79		Good	49.11
	Very good	27.58		Very good	28.08

In the survey sample, more than 80% of participants rated the air quality, landscaping conditions, and wetland protection status in their surrounding wetlands as good or very good. About 70% of respondents considered water quality and urban cleanliness to be good or very good. However, only 45.77% of participants rated biodiversity richness as good or very good, indicating that there is still room for improvement in biodiversity conservation in wetland management.

Overall, most participants believed that the likelihood of restoring wetlands after loss was small or impossible (Figure 2). About 68% of respondents thought the possibility of restoration was low, and 9% believed it was impossible. However, approximately 23% of participants believed that wetland restoration was possible, with 13% considering the likelihood to be high or very high. This suggests that some residents may lack understanding of wetland restoration efforts under current conservation practices (Figure 2A).

The majority of participants, about 82%, agreed that wetland protection is more important than economic development, with 16% strongly agreeing and 51% somewhat agreeing. However, approximately 14% of respondents disagreed with the idea that wetland protection is more important than economic development, and 4% strongly disagreed, indicating that some residents remain uncertain about the relationship between wetland conservation and economic growth (Figure 2B).

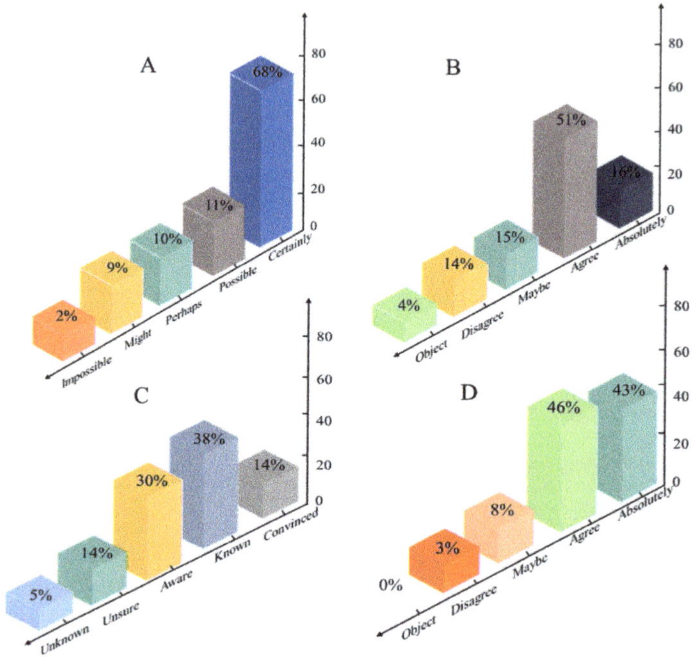

Figure 2. Residents' awareness of wetland conservation. Note: Panel (**A**) represents the likelihood of wetland restoration after loss. Panel (**B**) represents the belief that wetland protection is more important than economic development. Panel (**C**) represents the understanding of regulations related to wetland protection. Panel (**D**) represents support for the expansion of wetland areas in Beijing.

The survey also revealed that 19% of participants had little knowledge of wetland protection regulations, while 30% had a basic understanding. A further 38% and 13% of respondents reported being relatively well informed and very well informed, respectively, indicating that public education and outreach on wetland protection laws and policies still need to be strengthened (Figure 2C).

More than 90% of participants supported the expansion of wetland areas in Beijing, with only 3% opposing it. This shows that the ecosystem services and well-being provided by wetlands are widely recognized by the public (Figure 2D).

3.3. Residents' Behavior Toward Wetland Conservation

Beijing Wild Duck Lake Nature Reserve had the highest proportion of visitors (13.89%, Table 4). The most visited wetland park was the Beijing Cuihu Wetland Park, with a visitation rate of 23.02%. For drinking water source protection areas, the Miyun Reservoir had the highest visitation rate (10.52%). Among other protected areas, the top six most visited were Shichahai Wetland (37.20%), Summer Palace Wetland (36.71%), Chaoyang Park Wetland (33.83%), Olympic Forest Park Wetland (32.94%), Beihai Park Wetland (32.04%), and Yuanmingyuan Ruins Park Wetland (30.95%). These trends may be influenced by factors such as proximity to downtown Beijing, ease of transportation, and the cultural functions of the wetlands.

Table 4. Types of wetland protected areas visited by sample residents.

Indicator Explanation	Proportion (%)	Indicator Explanation	Proportion (%)
Wetland Nature Reserves (6)		Other Protected Areas (15)	
Beijing Wild Duck Lake Nature Reserve	13.89	Liulihe Wetland Forest Ruins Park	5.06
Hanshiqiao Wetland Nature Reserve	8.73	Shichahai Wetland	37.20
Huaihe River Water Wildlife Protection Area	3.08	Chaoyang Park Wetland	33.83
Juma River Aquatic Wildlife Protection Area	5.26	Yuanmingyuan Ruins Park Wetland	30.95
Baihepu Protection Area	4.86	Lotus Pond Park Wetland	13.10
Beijing Jinniu Lake Nature Reserve	5.26	Yongding River Fengtai Wetland	6.25
Wetland Parks (12)		Yongding River Shijingshan Wetland	7.44
Beijing Cuihu Wetland Park	23.02	Tongzhou Section of North CanalWetland	6.75
Beijing Wild Duck Lake National Wetland Park	15.08	Beijing Zoo Wetland	28.17
Beijing Changou Spring National Wetland Park	5.95	Summer Palace Wetland	36.71
Huairou Liulimiao Wetland Park	4.37	Yuyuantan Park Wetland	26.19
Daxing Changziying Wetland Park	7.14	Zizhuyuan Park Wetland	17.96
Nanhaizi Wetland Park	19.64	Beihai Park Wetland	32.04
Tanghekou Wetland Park	2.48	Olympic Forest Park Wetland	32.94
Pinggu Mafang Xiaolonghe Wetland Park	2.88	Taoranting Park Wetland	18.25
Yanggezhuang Wetland Park	1.79		
Mentougou Yanchi Jihe Wetland Park	3.27	Drinking Water Source Protection Areas (2)	
Miyun Reservoir	10.52	Chaobai River Upper Reach Wetland	1.88
Beijing Yuyuantan East Lake Wetland Park	10.42		
Miyun Mujiayu Hongmenchuan Wetland Park	1.88		

Most participants lived more than 2 km from the nearest wetland park, and only a small number lived within 500 m, indicating that most wetland parks are located away from residential areas (Figure 3). Overall, the number of visitors tended to increase as the distance from home to the wetland park increased. The majority of participants visited wetland parks at least once a month, with most visiting 1–3 times per month. However, some participants visited wetland parks less than once a month, a number second only to those who visited 1–3 times monthly, indicating that a portion of residents do not have a habit of visiting wetland parks regularly.

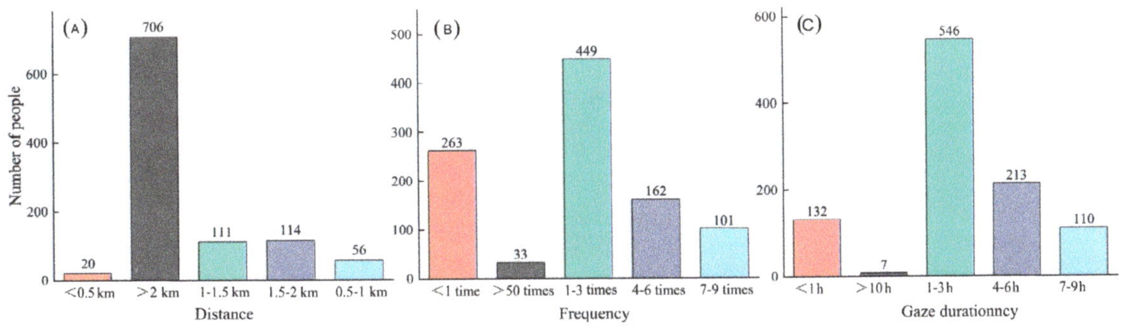

Figure 3. Residents' behaviors toward visiting wetland protected areas. Note: (**A**) represents the distance from residential areas to wetland parks, (**B**) represents the average frequency of visiting protected areas per month, and (**C**) represents the average time spent per visit in the protected area.

In terms of duration, most participants were willing to stay at wetland parks for more than 1 h per visit, with the majority spending 1–3 h at the parks. A smaller proportion spent 4–6 h, while some reported staying for less than 1 h per visit, suggesting that certain residents either lack interest in exploring wetland parks or find nearby parks too small for longer visits.

3.4. Residents' Willingness to Pay for Wetland Conservation

A total of 889 respondents indicated their willingness to support wetland conservation in Beijing through donations, far surpassing those who were unwilling to donate (Figure 4). This suggests that residents' awareness of wetland conservation has gradually increased, and they are willing to contribute financially to support and protect Beijing's wetlands. Overall, the number of participants willing to donate decreases as the maximum annual donation amount increases. Most respondents were willing to donate less than CNY 100, between CNY 100 and 299, and between CNY 300 and 499 per year for wetland protection, reflecting a possible financial constraint among residents.

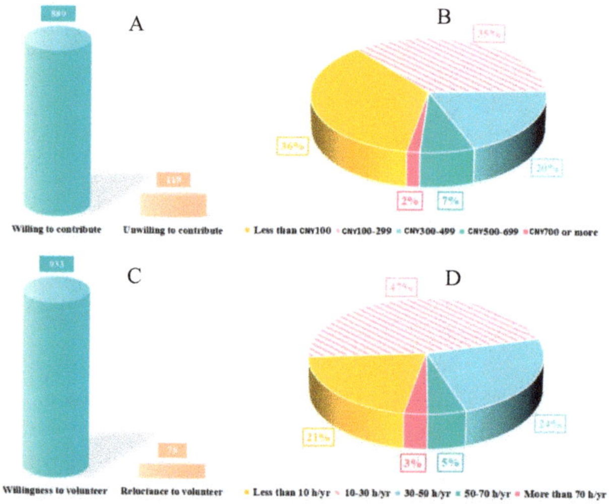

Figure 4. Residents' willingness to pay for wetland conservation. Note: (**A**) represents whether you are willing to contribute to the protection of Beijing wetlands through donations. (**B**) represents the maximum amount of money you are willing to donate annually for the protection of Beijing wetlands. (**C**) represents whether you are willing to support the protection of Beijing wetlands through volunteer work. (**D**) represents the maximum amount of time you are willing to volunteer annually for the protection of Beijing wetlands.

Additionally, 933 respondents expressed willingness to support wetland conservation through volunteer work, significantly more than the 75 who were unwilling. This highlights a growing awareness of wetland conservation among residents, who are eager to contribute through voluntary activities. In terms of time commitments, the majority of participants were willing to volunteer for 10–30 h (47%), 30–50 h (24%), and 50–70 h (5%) per year. Only 21% of participants were willing to volunteer for fewer than 10 h per year, indicating a high level of enthusiasm for participating in wetland conservation efforts.

4. Discussion

In this study, wetlands were categorized based on their management intensity, ranging from highly managed parks to less intensively managed nature reserves and reservoirs. However, each wetland's ecological characteristics, such as proximity to rivers and specific management types, were not explicitly considered in the survey analysis. Future studies should aim to classify wetlands more rigorously, considering ecological factors such as water quality, biodiversity, and access to nearby communities. Residents have a strong perception of changes in the wetland ecosystem, which is influenced by multiple internal and external factors [14]. Individual socio-economic characteristics are closely related to participants' awareness and attitudes toward current wetland conservation. Specifically, factors such as age, education level, marital status, and health status all have a significant impact on the evaluation results ($p < 0.05$). The survey shows that Beijing residents hold a positive attitude toward wetland conservation, with over 90% supporting the expansion of wetland areas, and more than 82% believing that urban wetland conservation is more important than economic development. This awareness mainly stems from residents' daily interactions with the cultural and recreational values of wetlands. However, despite high overall awareness, residents' understanding of the ecological functions of wetlands, such as water quality improvement and biodiversity conservation, remains relatively low. This could be attributed to the demographic characteristics of the respondents, as many were younger and had lower levels of education beyond undergraduate degrees.

Age plays an important role in shaping residents' perceptions, with younger respondents tending to focus on leisure and esthetic experiences, while older respondents generally have a deeper understanding of the potential ecosystem value [15]. Educational level is also a key factor influencing public awareness of wetland environments. For example, in a study on urban wetlands in Mexico City, education levels significantly affected residents' environmental awareness, with higher education correlating with stronger environmental perception [10]. Similarly, research in Nepal's Ghodaghodi Lake area found that higher education levels were associated with more positive conservation attitudes [16].

The geographic location and functional layout of wetlands significantly impact residents' visitation behavior. The duration and district of residence significantly affect participants' behavior toward wetland conservation ($p < 0.05$). Residents who are more closely connected to wetlands tend to have a more comprehensive understanding of the environment [8]. This study shows that wetlands near urban areas, such as Shichahai and Summer Palace Wetlands, had higher visitation rates than those in more remote areas like Huaihe River Water Wildlife Protection Area and Baihepu Protection Area. This suggests that accessibility and cultural services are key factors attracting residents to visit wetlands. Most residents visit wetlands 1–3 times a month, and the average duration of each visit is 1–3 h, reflecting the role of urban wetlands as important green spaces that meet residents' needs for ecological services and provide daily recreation and relaxation, effectively improving their quality of life and well-being.

In the future, the rational spatial layout of urban wetlands should be a key consideration in city planning. Additionally, urban wetland parks should be equipped with more educational and cultural facilities to enhance their ecological and cultural service functions, attracting more public participation in conservation-related activities.

Income level has a significant positive impact on both time investment and willingness to pay for wetland conservation (Figure 5). Income level significantly affects willingness to pay but has little impact on time investment, suggesting that higher economic capacity may drive fundraising for wetland conservation. Older respondents tend to have lower participation in wetland conservation, and specific guidance or incentive measures are needed to enhance their willingness to participate. Willingness to pay and actual participation in wetland conservation are important indicators of public awareness and pro-conservation behavior [17]. Personal variables, household income levels, environmental background, and social and human capital significantly influence residents' willingness to engage in wetland protection [18]. Although most residents are willing to donate to wetland conservation, their contributions are generally in lower financial brackets. This could be due to income constraints, as income levels have a positive effect on willingness to pay [19]. As income decreases, so does the public's willingness to contribute financially.

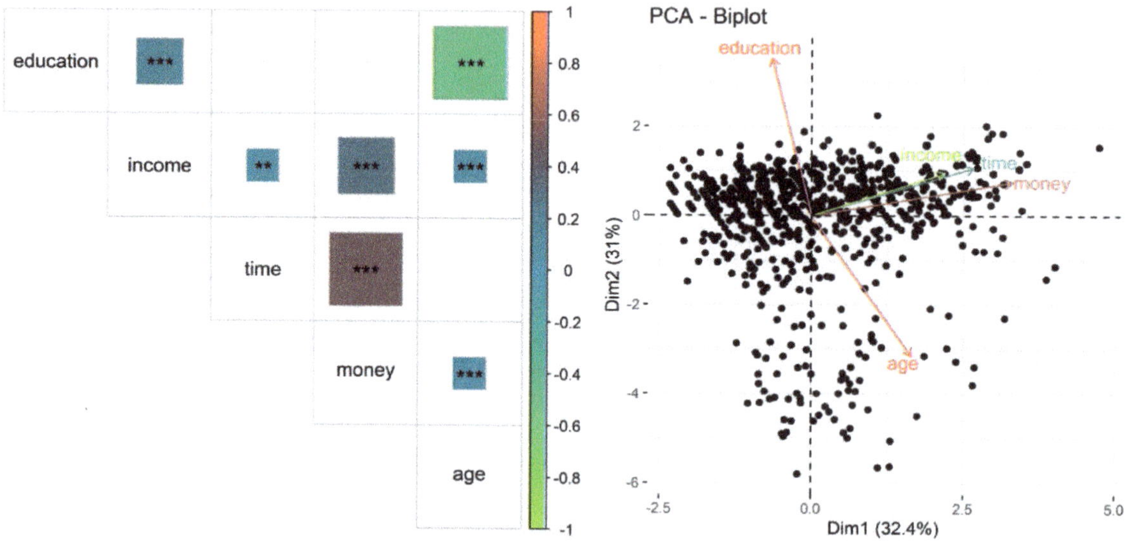

Figure 5. Willingness to pay and residents' characteristics. Note: the asterisks (*) are used to denote levels of statistical significance. ** stands for a significance level of $p < 0.01$. *** stands for a significance level of $p < 0.001$. In the matrix shown, the asterisks next to the pairs of variables are indicating the statistical significance of the relationships between those variables.

Moreover, distance from the residential area to wetlands also affects willingness to pay; residents living farther away from wetlands exhibit lower willingness to pay [20]. In contrast, residents' willingness to volunteer their time for conservation efforts tends to be higher than their willingness to donate money [21,22]. This study confirms this finding, as over 92% of residents expressed willingness to support wetland conservation through volunteer work. This indicates a high level of public interest in wetland protection. In the future, wetland management authorities should actively explore innovative public engagement mechanisms, such as seminars, lectures, and other fruitful activities, to encourage and build residents' capacities and awareness. These activities could help institutionalize public participation mechanisms and further strengthen residents' involvement in wetland conservation efforts.

5. Conclusions

This study provides an insightful analysis of Beijing residents' awareness, attitudes, and behaviors toward wetland conservation. The findings reveal that the majority of residents demonstrate a strong commitment to wetland protection, with over 90% supporting the expansion of wetland areas. Despite this positive attitude, this study also uncovers gaps in residents' understanding of specific ecological functions, such as biodiversity conservation and water quality improvement. Furthermore, while financial contributions for wetland conservation remain modest, a significant proportion of residents express a willingness to engage in volunteer efforts, indicating a high level of public interest in actively supporting conservation initiatives. This study highlights the importance of improving public education on wetland ecological services and suggests that the strategic spatial planning of urban wetlands, alongside enhanced cultural and educational facilities, could foster greater community engagement. Wetland management authorities should also focus on formalizing and institutionalizing public participation mechanisms to further strengthen residents' involvement in wetland conservation efforts. This comprehensive approach will be crucial for achieving sustainable wetland management in the long term. The findings of this study are based on data collected in Beijing, which may limit their applicability to other urban contexts. To address this, future studies should investigate whether similar patterns of awareness, attitudes, and behaviors are observed in cities with different wetland management practices, cultural values, and socio-economic conditions. Additionally, this study's reliance on self-reported data presents a potential limitation, as responses may be influenced by social desirability bias. Expanding future research to include a broader range of demographic groups and geographic regions could provide a more comprehensive understanding of public engagement in wetland conservation. Furthermore, longitudinal studies are essential to evaluate how public attitudes and behaviors change over time, particularly after the implementation of targeted educational initiatives or volunteer programs. Such research would offer valuable insights into the long-term effectiveness of public engagement strategies in promoting sustainable wetland management. The results of this study provide a valuable reference for policy-makers, urban planners, and conservationists seeking to integrate public engagement into wetland management strategies, and for fostering a greater collective effort in protecting Beijing's vital wetland ecosystems.

Author Contributions: L.M.: investigation, data curation. B.Y.: writing—review and editing. All authors have read and agreed to the published version of this manuscript.

Funding: This research was funded by the Fundamental Research Funds for the Central Universities (3142024041).

Data Availability Statement: The data presented in this study are available on request from the corresponding author.

Conflicts of Interest: The authors declare no conflicts of interest.

References

1. Russi, D.; Brink, P.T.; Badura, T. *The Economics of Ecosystems and Biodiversity for Water and Wetlands*; IEEP: London, UK; Brussels, Belgium, 2013.
2. Das, A.; Basu, T. Assessment of peri-urban wetland ecological degradation through importance-performance analysis (IPA): A study on Chatra Wetland, India. *Ecol. Indic.* **2020**, *114*, 106274. [CrossRef]
3. Ramachandra, T.V. Conservation and Management of Urban Wetlands: Strategies and Challenges. 2009. Available online: https://www.researchgate.net/publication/256932310_Conservation_and_Management_of_Urban_Wetlands_Strategies_and_Challenges (accessed on 10 October 2011).
4. Mo, L.C. Measurement and evolution analysis of the ecological service value of Beijing wetlands. *J. Ecol. Rural Environ.* **2024**, *40*, 749–756.
5. Mo, L.C.; Chen, J.C.; Xie, Y. Wetlands conservation in Beijing of China: Present status and development strategies. *Watershed Ecol. Environ.* **2024**, *5*, 73–79. [CrossRef]
6. Wang, Y.; Li, X.; Sun, M.; Yu, H. Managing urban ecological land as properties: Conceptual model, public perceptions, and willingness to pay. *Resour. Conserv. Recycl.* **2018**, *133*, 21–29. [CrossRef]
7. Gobster, P.H.; Nassauer, J.I.; Daniel, T.C.; Fry, G. The shared landscape: What does aesthetics have to do with ecology? *Landsc. Ecol.* **2007**, *22*, 959–972. [CrossRef]
8. Dobbie, M.; Green, R. Public perceptions of freshwater wetlands in Victoria, Australia. *Landsc. Urban Plan.* **2013**, *110*, 143–154. [CrossRef]
9. Trenholm, R.; Haider, W.; Lantz, V.; Knowler, D.; Haegeli, P. Landowner preferences for wetlands conservation programs in two Southern Ontario watersheds. *J. Environ. Manag.* **2017**, *200*, 6–21. [CrossRef] [PubMed]
10. Torres-Lima, P.; Conway-Gómez, K.; Buentello-Sánchez, R. Socio-environmental perception of an urban wetland and sustainability scenarios: A case study in Mexico City. *Wetlands* **2018**, *38*, 169–181. [CrossRef]
11. Truong, D.D. Villagers' Perception and Attitude Toward Wetland Values and Conservation in Vietnam: A Case Study of Xuan Thuy Ramsar National Park. *Front. Sociol.* **2021**, *6*, 763743. [CrossRef] [PubMed]
12. Hu, C.; Wright, A.L.; He, S. Public perception and willingness to pay for urban wetland ecosystem services: Evidence from China. *Wetlands* **2022**, *42*, 19. [CrossRef]
13. Zhang, L.S.; Yue, M.Y.; Qu, L.; Ren, B.; Zhu, T.; Zheng, R. The influence of public awareness on public participation in environmental governance: Empirical evidence in China. *Environ. Res. Commun.* **2024**, *6*, 1–20. [CrossRef]

14. Obradović, S.; Stojanović, V.; Milić, D. The importance of understanding local community attitudes and perceptions regarding nature conservation. *Wetlands* **2023**, *43*, 2. [CrossRef]
15. Yang, H.Z.; Li, H.M.; Zhang, A.L. Perception of pastoral households on grassland ecological degradation in the Sanjiangyuan area. *Arid Zone Res.* **2016**, *33*, 822–829.
16. Sah, J.P.; Heinen, J.T. Wetland resource use and conservation attitudes among indigenous and migrant peoples in Ghodaghodi Lake area, Nepal. *Environ. Conserv.* **2001**, *28*, 345–356. [CrossRef]
17. Alarcon, G.G.; Fantini, A.C.; Salvador, C.H.; Farley, J. Additionality is in detail: Farmers' choices regarding payment for ecosystem services programs in the Atlantic forest, Brazil. *J. Rural Stud.* **2017**, *54*, 177–186. [CrossRef]
18. Xu, X.P. Differences in public perception of wetland value and conservation: Based on a survey in Panshan County, Liaoning Province. *J. Hunan Agric. Univ. (Soc. Sci. Ed.)* **2014**, *15*, 95–100.
19. Cranford, M.; Mourato, S. Community conservation and a two-stage approach to payments for ecosystem services. *Ecol. Econ.* **2011**, *71*, 89–98. [CrossRef]
20. Hanink, D.M. The economic geography in environmental issues: A spatial-analytic approach. *Prog. Hum. Geogr.* **1995**, *19*, 372–387. [CrossRef]
21. Dai, H.X.; Li, J.H.; Cheng, K.; Zeng, C. Willingness of community residents in the Sanjiang Plain reserve to contribute to wetland ecosystem services. *J. Nat. Resour.* **2017**, *32*, 977–987.
22. Brinkmann, K.; Hoffmann, E.; Buerkert, A. Spatial and temporal dynamics of urban wetlands in an Indian megacity over the past 50 years. *Remote Sens.* **2020**, *12*, 662. [CrossRef]

water

MDPI

Article

Impact of Ecological Restoration on Carbon Sink Function in Coastal Wetlands: A Review

Xiaoqun Guo [1,2,3], Yanjin Liu [1,2,4], Tian Xie [4], Yina Li [4], Hongxi Liu [1,2,4] and Qing Wang [1,2,4,*]

1 Research and Development Center for Watershed Environmental Eco-Engineering, Advanced Institute of Natural Sciences, Beijing Normal University, Zhuhai 519087, China; 13406747521@163.com (X.G.); 202431180059@mail.bnu.edu.cn (Y.L.); liuhongxi@bnu.edu.cn (H.L.)
2 Key Laboratory of Coastal Water Environmental Management and Water Ecological Restoration of Guangdong Higher Education Institutes, Beijing Normal University, Zhuhai 519087, China
3 College of Geography and Environmental Science, Northwest Normal University, Lanzhou 730071, China
4 State Key Laboratory of Wetland Conservation and Restoration, School of Environment, Beijing Normal University, Beijing 100875, China; tianxie@bnu.edu.cn (T.X.); 202131180022@mail.bnu.edu.cn (Y.L.)
* Correspondence: wq@bnu.edu.cn

Abstract: Reducing carbon emissions and increasing carbon sinks have become the core issues of the international community. Although coastal blue carbon ecosystems (such as mangroves, seagrass beds, coastal salt marshes and large algae) account for less than 0.5% of the seafloor area, they contain more than 50% of marine carbon reserves, occupying an important position in the global carbon cycle. However, with the rapid development of the economy and the continuous expansion of human activities, coastal wetlands have suffered serious damage, and their carbon sequestration capacity has been greatly limited. Ecological restoration has emerged as a key measure to reverse this trend. Through a series of measures, including restoring the hydrological conditions of damaged wetlands, cultivating suitable plant species, effectively managing invasive species and rebuilding habitats, ecological restoration is committed to restoring the ecological functions of wetlands and increasing their ecological service value. Therefore, this paper first reviews the research status and influencing factors of coastal wetland carbon sinks, discusses the objectives, types and measures of various coastal wetland ecological restoration projects, analyzes the impact of these ecological restoration projects on wetland carbon sink function, and proposes suggestions for incorporating carbon sink enhancement into wetland ecological restoration.

Keywords: coastal wetland; ecological restoration; carbon sequestration capacity

1. Introduction

A series of ecological and environmental problems caused by global climate change have become major threats to the sustainable development of human society and are among the greatest challenges worldwide. Carbon reduction and carbon sink enhancement have become key issues and strategies of the international community [1]. Oceans, including coastal ecosystems, constitute the largest active carbon pool on Earth, and they have great potential to sequester carbon within sediment or organisms [2,3]. According to the Sixth Assessment Report of the IPCC, the number of marine carbon sinks is significantly greater than the number of terrestrial carbon sinks. Coastal carbon, as a special type of carbon between marine carbon and terrestrial carbon, is a carbon pool that humans can affect and regulate [4]. In the 2019 Special Report on Oceans and Cryosphere in a Changing Climate, mangroves, seagrass beds, coastal salt marshes and macroalgae were listed as four major coastal blue carbon ecosystems. Ecosystems such as mangrove and seagrass beds in coastal wetlands cover less than 0.5% of the seabed, but their carbon reserves account for more than 50% of the marine carbon reserves, making them important components of the blue carbon ecosystem [5]. Coastal blue carbon ecosystems, such as mangroves and salt marshes, play an important role in mitigating global climate change because of their ability to preserve carbon from land, ocean and atmosphere in their living organisms (biomass) and sediments [6–8]. Consequently, the function and mechanisms of carbon sinks have become important topics worldwide, especially in the face of climate change and human disturbances.

Coastal wetlands lie in the transition zone between land and sea. They are not only important carbon sinks but also provide various ecosystem functions and services, such as coastal defense [9–11], productivity and biodiversity maintenance [12,13] and water purification [14,15]. The blue carbon in coastal areas refers to the carbon dioxide extracted from the atmosphere, fixed by the interaction of plants and microorganisms, and stored in nearshore sediments and soils through the action of higher plants (including salt marshes, plants, mangroves, and seagrasses), phytoplankton, macroalgae, and marine calcifiers, as well as the carbon transported from coastal areas to the ocean and seabed [16]. This carbon storage process is not only highly important to the global carbon cycle but also plays a crucial role in global climate change mitigation strategies. For example, in North Carolina, USA, a coastal salt marsh wetland with an area of only 0.25 km^2 has an annual carbon burial amount equivalent to the amount of carbon dioxide emitted from burning 28,000 L of gasoline [17]. Furthermore, the carbon in coastal blue carbon ecosystems is stored in the soil, aboveground living biomass (leaves, branches, trunks), belowground living biomass (roots), and nonliving biomass (such as litter and dead wood) of mangroves, salt marshes, and seagrass beds [18]. Like the carbon stored in terrestrial ecosystems, blue carbon also includes carbon that is fixed by living plants over a relatively short period (a few years to several decades). However, the carbon sequestration per unit area in coastal blue carbon ecosystems is much greater than that in terrestrial carbon sinks. The annual carbon sequestration rate of coastal blue carbon ecosystems reaches 2.2×10^4 Mg CO_2 ha^{-1}, which is dozens to hundreds of times greater than that of terrestrial forest ecosystems [19]. Moreover, because coastal wetlands have high sedimentation rates and low organic carbon decomposition rates, they have great potential as carbon sinks [20,21]. The carbon sinks of coastal wetland ecosystems consist mainly of plant carbon pools and sediment carbon pools. A plant carbon pool is a carbon pool formed by carbon fixation through photosynthesis in the ecosystem, i.e., a biomass carbon pool, which usually changes with the growth of plants [22]. A sediment carbon pool can be divided into endogenous carbon and exogenous carbon according to its source. Endogenous carbon is the carbon formed by plants themselves and enters the sediment, whereas exogenous carbon is carbon from adjacent ecosystems (such as river basins and nearshore waters) that is transported by sediments in water bodies and is captured, buried and added to the sediment carbon pool. Specifically, there is interaction between the sediment carbon pool and the plant carbon pool [22]. Obviously, the ecosystem carbon sink function is influenced by the factors driving the vegetation carbon pool and sediment carbon pool, which is a multifactor process.

Since the beginning of the 21st century, with the continuous development of the economy and the expansion of human activities, global coastal wetlands have deteriorated, and the carbon sequestration capacity of wetlands has been severely restricted [23–25]. Due to urbanization, development and global changes, coastal wetlands in many parts of the world have been lost or degraded [26,27]. From 1980 to 2000, approximately 50% of the world's tidal marshes, 35% of mangroves, 29% of seagrass beds and 30% of coral reefs disappeared or degraded [9,28]. From 1984 to 2016, approximately 16% of tidal flats disappeared [29]. The massive loss and degradation of coastal wetlands have made coastal wetlands among the most endangered ecosystems in the world. Notably, the degradation of natural wetlands leads to the release of a large amount of previously stored soil carbon in the form of carbon dioxide (CO_2) and, to a lesser extent, methane (CH_4), effectively transforming natural wetlands from carbon sinks to carbon sources [30,31].

This immense loss and degradation have made coastal wetlands among the most endangered ecosystems in the world, and ecological restoration is a primary means of enhancing biodiversity resilience and restoring fundamental ecosystem services [32,33]. At present, international wetland restoration methods focus mainly on the restoration of hydrological conditions in damaged wetlands, the breeding of halophytes and salt-tolerant plants, and the restoration of species and habitats [34]. For example, through the planting of aquatic plants, a lake built for the deposition of rare metal ores near Capel in Australia has been restored to a healthy wetland ecosystem [35]. As ecological restoration has become a vital strategy for the protection and management of China's future coastal wetlands, an increasing number of restoration projects will be implemented with government policy and financial support (China National Wetland Conservation Action Plan, 2022). Some wetland ecological restoration projects have already shown significant effects. For example, a restoration project in the Yellow River Delta of China has expanded the area of reed wetlands, increasing the diversity and habitat quality of waterfowl [36]. A restoration project in Chongming Dongtan, China, used topographical transformation, water system construction, and vegetation management to create habitats for waterfowl in areas where the invasive Spartina alterniflora had been removed, significantly increasing the diversity of waterfowl [37]. Although the implementation of wetland restoration projects has made some progress, we still need to conduct more in-depth discussion and research on the impact of ecological restoration measures on the carbon sequestration function of wetlands.

Importantly, coastal wetland restoration might also enhance the carbon sink function, and such projects should be characterized by high safety, stability and efficiency [38]. However, because of the variety of restoration projects that have occurred at wetland sites, with differences in scale, measure, goal and object, the specific effects of restoration efforts on wetland carbon sink function differ. Furthermore, many restoration projects aimed at enhancing habitat restoration and biodiversity might not consider the carbon sink function. Moreover, the effects of measures used in restoration projects (i.e., hydrological regulation or replenishment) on the carbon sink function have not been well considered.

Since wetland carbon sinks are affected by various factors, the specific relationships between wetland carbon sinks and various ecological restoration projects need to be described in more detail. To comprehensively discuss this issue, we review the research status and factors influencing coastal wetland carbon sinks, discuss the goals, types and measures of various coastal wetland ecological restoration projects, and analyze the impacts of various ecological restoration projects on wetland carbon sink function. Finally, we propose implications for wetland ecological restoration efforts that enhance the carbon sink function. We hope to outline the key factors of different types of ecological restoration projects and the effects of these projects on wetland carbon sequestration and then apply this information to the design and management of restoration projects in coastal wetlands to maximize the benefits of ecological restoration to the carbon sequestration capacity in the face of ongoing environmental challenges.

2. Coastal Wetland Carbon Sink Function and Its Influencing Factors

The blue carbon function of coastal wetlands is the result of the interaction and balance between multiple processes, such as photosynthetic carbon absorption, carbon deposition and burial, and carbon export (Figure 1). The key processes mainly include the following: (1) Vegetation photosynthetic carbon fixation. Plants in coastal wetlands absorb atmospheric carbon dioxide (CO_2) through photosynthesis in a process known as photosynthetic carbon fixation. (2) Photosynthetic carbon allocation. The transfer pattern of photosynthetic carbon to the root system and soil carbon pool is key to the formation of the sedimentary carbon pool [39]. (3) Carbon deposition and burial. The remaining balance between carbon fixed by photosynthesis and carbon emitted as greenhouse gases is rapidly buried by sediments in a process known as sedimentary carbon fixation. This is reflected mainly vertically through sediment carbon burial rates and horizontally through tidal action with the exchange of dissolved inorganic carbon (DIC), dissolved organic carbon (DOC), and particulate organic carbon (POC) in seawater. (4) Soil carbon loss. Tidal action leads to intermittent anaerobic and aerobic alternations in wetlands, frequent redox processes, and interactive effects between terrestrial and marine microbes; at the same time, periodic tides carry a large amount of SO_4^{2-}, which hinders CH_4 production, leading to soil carbon loss [40].

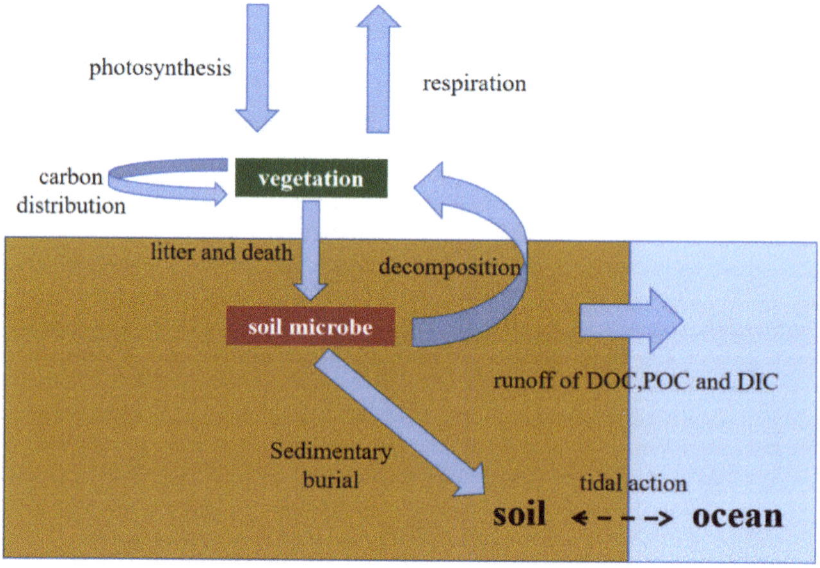

Figure 1. The key processes of carbon sequestration in coastal wetlands under tidal action include vegetation photosynthetic carbon fixation, photosynthetic carbon allocation, carbon deposition and burial, soil carbon mineralization and decomposition, and soil carbon loss.

Plants are important media for the exchange of matter and energy across the entire terrestrial surface, serving as the primary carbon reservoir [41]. They can transform atmospheric CO_2 into carbohydrates through photosynthesis; therefore, plant photosynthetic carbon fixation is the starting point for carbon sequestration in coastal wetlands.

Photosynthetically fixed carbon is allocated to different parts of plants, and the pattern of carbon allocation directly affects the direction of the sediment carbon pool. This process is controlled mainly by nonbiological factors such as sediment pore water salinity and nutrients. An in-depth analysis of the mechanisms of photosynthetic carbon allocation in plant–soil systems is a prerequisite for estimating the carbon sequestration function of coastal wetlands. Plant photosynthetic carbon allocation is an adaptive strategy employed by plants to cope with environmental stress [21,42]. Climate conditions, soil physicochemical properties, microbial activity and plant growth and development affect the patterns of plant photosynthetic carbon allocation, altering the carbon input to aboveground and belowground parts [43]. For example, as plant-available nutrients become increasingly limited under drought stress, plants may allocate more photosynthetic carbon to roots to absorb water, thereby triggering soil organic matter decomposition and increasing the availability of plant nutrients [44]. Moreover, drought can accelerate or decelerate soil carbon cycling by regulating the quality and quantity of rhizosphere secretions [45]. Correspondingly, the hydrological processes of coastal wetland ecosystems are influenced mainly by periodic tidal action and the interaction between surface freshwater and groundwater, leading to soil salinization or surface flooding in wetlands [23], which may also change the strategies of plant photosynthetic carbon allocation. Furthermore, photosynthetic carbon fixation is the main driving force for carbon allocation, and environmental stress can also indirectly affect carbon absorption by influencing photosynthesis, thereby indirectly affecting the allocation of photoassimilates [46].

Coastal wetlands, due to their relatively high carbon sequestration and burial rates, have become an important component of the blue carbon sink function in these ecosystems, which mainly includes autochthonous and allochthonous organic carbon [18]. Autochthonous organic carbon refers to the accumulation of wetland vegetation detritus, whereas allochthonous organic carbon refers to the deposition of suspended particulate organic carbon from seawater during tidal inundation, including various algal and other biological necromass in the seawater and various forms of particulate organic matter transported from upstream rivers [47]. A portion of the organic matter that accumulates in coastal wetlands is degraded and mineralized by microorganisms to produce CO_2 and CH_4, while the other part is buried in sediments, becoming an important component of the carbon sequestration function of coastal wetlands [48,49]. The rate of carbon burial in coastal wetlands varies significantly depending on different vegetation types and is also affected by the rate of sea-level rise [50].

Soil carbon loss in coastal wetlands is closely related to marine tides, surface runoff and the activities of the soil biota [51]. Under the influence of hydrological processes such as marine tides and surface runoff, soil organic carbon enters adjacent waters in a dissolved form, which is an important pathway for carbon loss in coastal wetlands [52]. In addition to hydrological actions, the mineralization and decomposition of soil carbon, which occur with the participation of microorganisms, constitute a key pathway for carbon loss [53,54]. This includes the decomposition of plant litter, hydrolysis of soil organic carbon, fermentation, acid production and the generation of CO_2 and CH_4 [55]. Carbon sequestration in coastal wetlands is typically characterized by the accumulation of organic carbon in soil or sediments [56]. Understanding the forms of soil carbon and its synergistic relationship with microorganisms is fundamental to understanding the stability of soil carbon pools [57]. The high carbon retention capacity endows coastal wetlands with significant blue carbon sink functions [58]. Influenced by marine tides, surface runoff and soil biological activities, minor changes in soil carbon pools can greatly regulate the carbon balance of ecosystems, becoming an important factor in constraining the carbon sequestration potential of salt marsh wetlands [59]. The carbon sink function of coastal wetlands is comprehensively affected by many factors, and its complex processes and mechanisms are as described above. However, as one of the focuses of wetland ecological restoration research, it is necessary to study the shortcomings and limitations of the existing wetland carbon sequestration monitoring and calculation technology. This not only helps to accurately assess the current status of wetland carbon sinks, but also provides a scientific basis for the planning and implementation of subsequent ecological restoration projects. At present, although some achievements have been made in this field, there are still many problems to be solved at the technical level.

3. Status of Various Coastal Wetland Ecological Restoration Projects

Many coastal wetlands around the world are facing serious threats. Countries have launched ecological restoration efforts to address this challenge [60]. In Louisiana, USA, the restoration of the Mississippi River Delta wetlands is an important measure to combat wetland degradation, land loss and climate change [61]. Among them, the Mid-Barataria Sediment Diversion project aims to restore and maintain up to 40 square miles of wetlands over the next 50 years by constructing permeable gates on the Mississippi River levee to divert sediment into the Barataria Basin. In addition, the Florida Everglades, the largest subtropical wetland ecosystem in the United States, has been under restoration since the 1980s [62]. The Comprehensive Everglades Restoration Plan (CERP), passed in 2000, aims to protect biodiversity through measures such as restoring natural hydrological conditions, improving water quality, restoring habitats, and controlling invasive species. In recent years, CERP has made significant progress in water quality improvement and invasive species control, but it also faces challenges such as climate change, funding, and ecosystem complexity [63].

The ecological restoration of the Solway Firth wetlands in the UK enhances the functions of the wetland ecosystem and biodiversity through multiple measures, including establishing protected areas, restoring hydrological conditions, rehabilitating habitats, controlling invasive species, and strengthening community participation [64]. The close cooperation among government agencies, environmental organizations, and local communities has promoted the restoration of the wetland ecosystem and provided opportunities for eco-tourism and education for local residents and tourists [65]. The core of wetland restoration work is to restore wetlands disturbed or altered by human activities to their natural state. This includes rebuilding hydrological conditions and ecosystems while paying attention to maintaining the balance of biodiversity, temperature and vegetation within the wetlands, following natural and ecological laws. Due to the unique ecological characteristics of different types of wetlands such as swamps, mangroves, peatlands and salt marshes, their restoration strategies must also vary according to specific features, landscapes, hydrological connections and surrounding human activities [66]. For example, wetland restoration projects in Odisha, India, focused on establishing the Chilika Lake Wildlife Sanctuary and the Bharatpur Bird Sanctuary in Rajasthan, providing important habitats and breeding grounds for various bird species [67]. In addition, the Veppuram-Kol wetland ecological restoration project in Kerala, India, has strengthened the management of invasive fish species to control the damage of invasive species to the ecosystem.

Take China as an example. China's coastal wetlands cover an area of approximately 5.8 million hectares. However, in recent decades, due to urban expansion and the increasing demand for land from industrial and agricultural development, land reclamation has intensified, leading to a sharp decrease in coastal wetlands. The areas of intertidal marshes, tidal flats, mangroves and coral reefs decreased by 57%, 73% and 80%, respectively, from the 1950s to the 2010s, and the area of seagrass beds decreased by more than 80% [67]. The invasion and rapid spread of the exotic species Spartina alterniflora, which reached 51,800 hectares in 2018, is one of the main reasons for the degradation of China's coastal wetlands [68]. To protect coastal wetlands, the Chinese government invested over USD 760 million between 2006 and 2015, supporting hundreds of coastal wetland restoration projects [69]. For example, a restoration project in the Yellow River Delta expanded the area of reed wetlands, increasing the diversity and habitat quality of waterfowl [36]. A restoration project in Chongming Dongtan created habitats for waterfowl in areas where the invasive Spartina alterniflora had been removed, using topographical transformation, water system construction and vegetation management, which greatly improved the diversity of waterfowl [37]. We collected data via a standardized search of the China Knowledge

Resource Integrated Database (CNKI) and Web of Science. The search terms were (coastal OR salt marsh OR mangrove OR coral reef OR seagrass bed) AND (restor* OR recovery OR rehab*) AND (China). A total of 86 restoration projects were identified (Figure 2). There were 29 projects aimed at enhancing species diversity, 22 projects aimed at improving ecosystem function, 38 projects aimed at improving habitat quality and 10 projects aimed at improving vegetation restoration (Figure 2a).

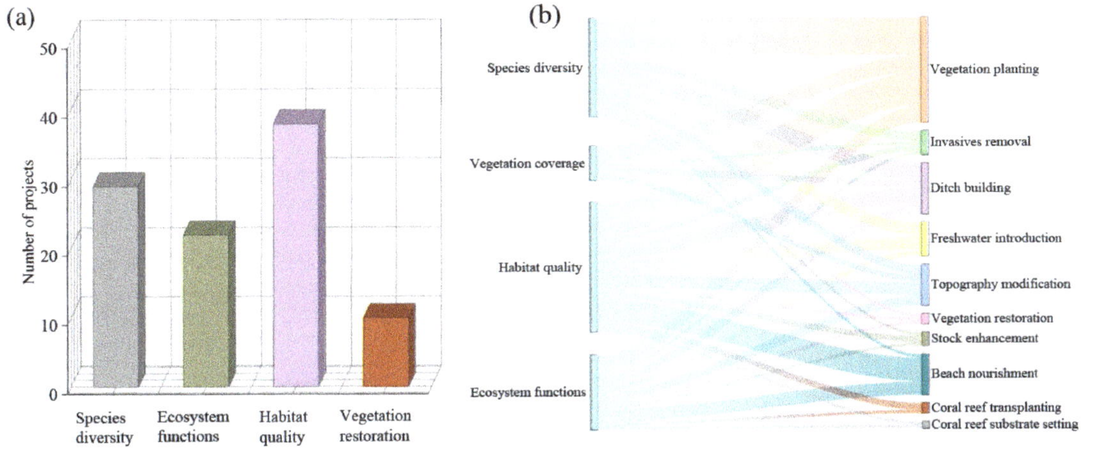

Figure 2. The bar chart represents the number of projects corresponding to different objectives of coastal wetland projects in China. (**a**) The horizontal axis lists four types of ecological restoration projects: species diversity, ecosystem functions, habitat quality and vegetation restoration. The vertical axis represents the number of items. Sankey figure represents the correlation analysis between the purpose of the ecological restoration project and the means of ecological restoration. (**b**) The left side lists four ecological restoration goals: species diversity, vegetation coverage, habitat quality and ecosystem functions. Various ecological restoration measures are listed on the right: vegetation planting, invasive species removal, ditch building, freshwater introduction, topography modification, vegetation restoration, stock enhancement, beach nourishment, coral reef transplanting and coral reef substrate setting.

In wetland restoration projects, improving habitat quality and enhancing species diversity are the core objectives, and in order to effectively achieve the objectives, 21 projects have chosen "vegetation planting" as a means of restoration (Figure 2b). Vegetation planting can not only quickly cover bare land and reduce soil erosion but also provide the necessary habitat and food sources for various organisms, thereby directly contributing to the increase in species diversity [70,71]. By carefully selecting and planting native species, we can ensure that vegetation is highly adaptable to the local ecological environment and increase the stability and resistance of the ecosystem [21,72]. Moreover, the lush growth of vegetation helps regulate the local climate, reduce temperature, increase humidity and provide a more comfortable living environment for organisms [73]. Therefore, as an economical, efficient and environmentally friendly restoration technology, vegetation planting has been widely used in various wetland restoration projects, aiming to achieve long-term health and sustainable development of the ecosystem by restoring and enhancing the natural ecological functions of wetlands [74]. In addition, five projects used a combination of "invasive removal" and "vegetation planting" to provide a comprehensive and effective ecological restoration strategy for wetland restoration projects [75]. This strategy focuses not only on addressing current ecological issues but also on the long-term health and sustainable development of the ecosystem, ensuring that wetlands can provide ecological services such as biodiversity conservation, soil and water conservation and water purification while adapting to future environmental challenges [76].

Moreover, a total of 25 projects used restoration techniques including freshwater introduction and ditch building. The main purpose of these technologies is to improve and simulate natural hydrological conditions in order to promote the overall health of ecosystems and improve the quality of biological habitats [66,77]. By introducing fresh water, the salinity of wetlands can be reduced, providing a more suitable environment for freshwater-dependent species [78]. Moreover, hydrological regulation technology, such as the construction of gates, pumping stations or regulating reservoirs, can effectively manage the water level and flow of wetlands and maintain the hydrological balance of wetlands [79,80]. The comprehensive application of these technologies can not only restore the natural hydrological characteristics of wetlands [81] but also improve the adaptability of wetlands to climate change and ensure the long-term stability and health of wetland ecosystems [82].

A total of 12 project restoration methods included topographic transformation, which changed the land use type, such as the conversion of farmland/culture ponds. This strategy restores wetlands to their natural state by stopping or reducing agricultural activities, thereby protecting and increasing biodiversity and providing necessary habitats for rare and endangered species [83]. In addition, the restoration of natural wetlands helps to improve the carbon storage capacity of wetlands, combat global climate change [84] and enhance the self-purification capacity [85], flood storage and climate regulation function of wetlands by restoring their natural cycles and biogeochemical processes [86].

4. Impact of Various Restoration Projects on the Carbon Sink Function in Coastal Wetlands

After an in-depth analysis of various coastal wetland ecological restoration projects, we clearly recognize that these restoration projects play important roles in protecting and restoring coastal wetland ecosystems. There are significant differences in the methods adopted by different types of restoration projects, and these differences will undoubtedly have a profound impact on the carbon sink function of coastal wetlands (Figure 3).

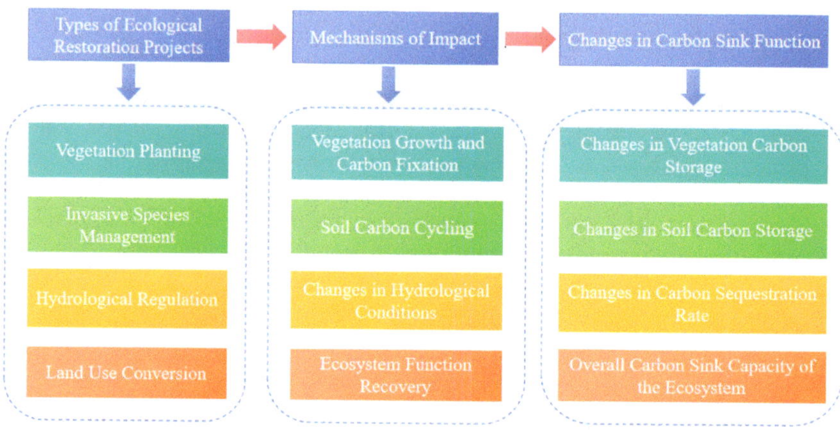

Figure 3. Conceptual model of the impact of ecological restoration projects on the carbon sink function of coastal wetlands. The blue arrows represent the specific content of ecological restoration project type, impact mechanism and carbon sink function change, the red arrows represent the impact process of ecological restoration project on the carbon sink function of coastal wetland and the boxes of the same color represent the corresponding impact process.

They induce changes in land types and represent different carbon sequestration processes [87,88]. We collected carbon density data for different vegetation types in the coastal provinces of Liaoning, Hebei, Tianjin, Shandong, Jiangsu, Shanghai, Zhejiang, Fujian, Taiwan, Guangdong, Guangxi and Hainan to produce the graph below. It can be seen from the figure that soil organic carbon density varies with vegetation coverage and location (Figure 4). Different restoration measures will lead to different carbon sequestration of soil and vegetation.

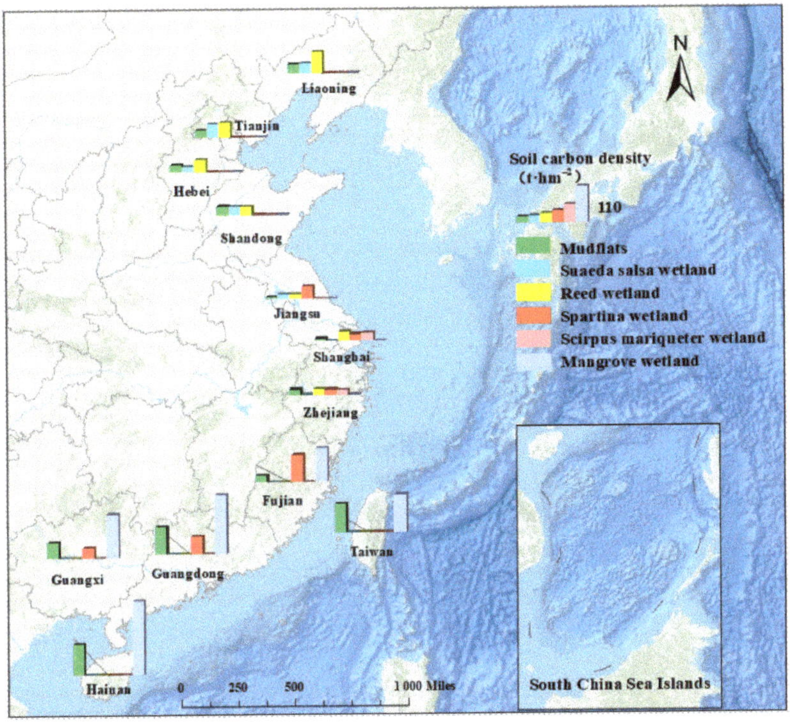

Figure 4. Soil carbon density of different wetland types in coastal areas of China (t·hm^{-2}). The histogram with different colors shows the soil carbon density of six wetland types, which are mudflats, *Suaeda salsa*, reed, *Spartina alterniflora*, *scirpus mariqueter* and mangrove wetland.

On this basis, this paper further explores how specific restoration measures, such as plant planting, invasive plant management, hydrological regulation and the conversion of farmland/culture ponds to wetlands, affect the carbon sequestration capacity of wetlands and proposes feasible guidance to provide more targeted and effective strategies for the ecological restoration of coastal wetlands and the improvement of carbon sink functions.

4.1. Vegetation Plantation

In wetland ecological restoration, the planting of suitable plant species can significantly increase the carbon sequestration capacity of wetlands [84]. A successful vegetation plantation will cause a transformation of land type, usually from a bare or low vegetation-covered state transforming to a vegetation-covered state (Figure 5), which will bring increased carbon sequestration both from restored vegetation (in biomass) and soil (in organic matter input). Specifically, within restoration implementation, trends of carbon sink function might be varied due to different measures. It mainly depends on the degree of disturbance to the soil, which fixes a lot of carbon and is released when the deeply buried soil is turned out [22]. The increment in carbon sink function is decided by the pre-restoration and post-restoration states of the land, while the latter is mainly influenced by restored vegetation species, growth condition (biomass), density and recovery time, etc.

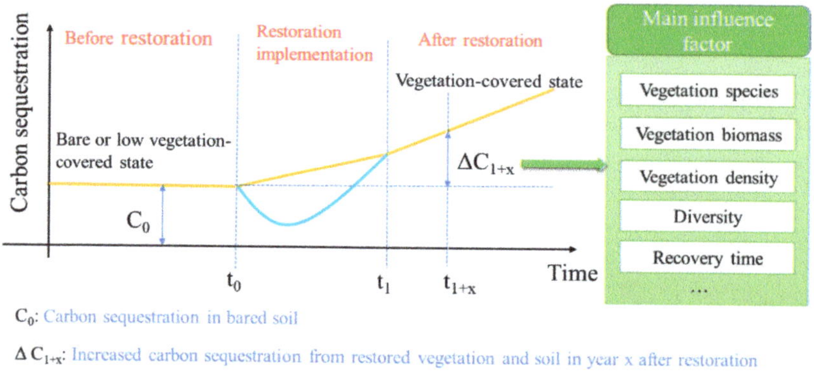

Figure 5. Impact of vegetation plantation on carbon sink function in coastal wetland. The orange and blue lines represent the two potentials of carbon sink function variation during restoration implementation. One potential carbon sink function is steadily increasing, with little disturbance to the soil. The other is decreasing first and then increasing due to a relatively large disturbance to the soil, which induces the release of carbon dioxide from the soil.

The carbon fixation capacity of different vegetation types varies due to a variety of factors, including the physiological structure and ecological characteristics of the plant itself, climatic conditions and human disturbance [89–91]. These factors collectively determine a plant's adaptability to environmental conditions, photosynthetic efficiency, growth rate and metabolic activities, which in turn affect the accumulation of plant biomass and carbon storage capacity. For example, genetic differences among plant species, symbiotic relationships with other organisms and the content of organic matter and microbial activity in the soil can all impact plant growth and carbon cycling [92]. Therefore, the differences in biomass, carbon density and carbon fixation capacity among various vegetation types reflect their distinct roles in the carbon cycle of ecosystems [93], which is crucial for understanding and managing the carbon sequestration function of wetland ecosystems. For example, the reed communities in Chongming Dongtan have greater biomass (6.28~11.74 kg/m^2) and carbon density (2.81~5.23 kg/m^2), as well as a greater carbon fixation capacity (1.24~2.02 kg·m^{-2}·a^{-1}), than the sea club-rush communities in the same area, which have relatively lower biomass (0.80~1.50 kg/m^2), carbon density (0.36~0.66 kg/m^2) and carbon fixation capacity (0.35~0.91 kg·m^{-2}·a^{-1}) [94].

These data indicate that by selecting appropriate species for planting, the carbon sequestration capacity of wetlands can be effectively enhanced. Specifically, planting species with high biomass and carbon density, such as reeds, which have strong carbon fixation capabilities, can lead to greater carbon storage in wetland ecosystems [95]. This is because such plants absorb more carbon dioxide through photosynthesis and convert it into organic carbon stored within the plant and soil [96]. Moreover, the slow decomposition of the roots and litter of these plants in the soil contributes to long-term carbon storage, thereby increasing the carbon sequestration capacity of wetlands [97]. Increasing the diversity of wetland plants can increase the stability and carbon sequestration capacity of the ecosystem [98]. Different plant species can form complementary relationships, which jointly improve the adaptability of the ecosystem to environmental changes [99]. Diverse plant communities can provide more niches, increasing the complexity of the ecosystem and thus enhancing its capacity to absorb and store carbon [100]. Lastly, recovery time is also important to carbon sequestration, as it decides the accumulated carbon in terms of the biomass increment of vegetation and organic matter input to soil per year.

4.2. Invasive Plant Control

For many ecosystems, the control of invasive plants is a common measure in wetland ecological restoration, such as the control of *Spartina alterniflora* in China. It was introduced in 1979 to protect the coastline. The strong adaptability has made it grow rapidly from 4375 hectares in 1990 to 68,000 hectares in 2022, becoming the most widely distributed dominant vegetation in the intertidal zone of coastal wetlands in China [5]. It has caused a series of problems, such as changing the habitat environment of offshore organisms, affecting the ability of beach aquaculture and sea water exchange, causing the blockage of harbor roads, etc., and then posing a threat to the health of coastal wetland ecosystem. However, it has a significant performance in carbon storage, which makes the effect of its exclusion on the carbon sequestration function of wetland complex (Figure 6). For many restoration projects, the control of invasive plants is a first step, while the latter steps include the recovery of native species. For example, the large-scale ecological restoration project in Chongming Dongtan, China, targeted the management of 24 km^2 of invasive *Spartina alterniflora*, aiming to eradicate invasive species and restore native vegetation. In this way, the carbon sequestration function of wetlands may be decreased firstly and then increased (Figure 6), and the carbon sequestration function of post-restoration is also decided by vegetation species, biomass, density, diversity and recovery time. The question is whether the carbon sequestration function can exceed the previous state with the invasive plants covered? This might be complex, which depends on the comparison of carbon sequestration capacity between the invasive and native species to some extent.

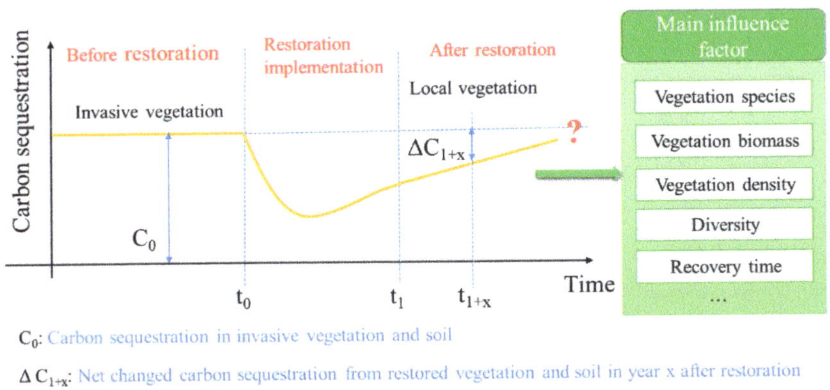

Figure 6. Impact of invasive vegetation control on carbon sink function in coastal wetland. The orange line represents the potential of carbon sequestration variation in pre-, in- and post-restoration. The decreasing trend of the orange line is caused by the exclusion of invasive vegetation and the disturbance to the soil, and the latter will induce the release of carbon dioxide from the soil. After the exclusion of invasive vegetation, native vegetation will be re-established or restored, which will cause an increase in carbon sequestration. However, the variation in the net changed carbon sequestration is uncertain and depends on the carbon sequestration capacity of the recovered vegetation, recovery time, etc.

The invasion of *Spartina alterniflora* has significantly affected interspecific competition among native plants, thereby causing complex changes in wetland carbon sequestration functions [101]. *Spartina alterniflora* does not directly invade salt marshes already covered by native vegetation but instead establishes and expands on mudflats, reducing the input of seawater salts and indirectly altering the competitive relationships among native plants [102]. The invasion of *Spartina alterniflora* has significantly increased the blue carbon storage capacity of salt marshes, which is related mainly to its expansion into mudflats and its efficient carbon sequestration capacity [103]. *Spartina alterniflora* possesses strong reproductive and dispersal abilities, with vigorous plant growth, a relatively long growing season, a relatively high leaf area index and relatively high net photosynthesis rates, thus resulting in relatively high primary productivity and biomass [104]. This leads to a significantly higher carbon storage than that of native species in some areas. For instance, its carbon storage was 16.9 times that of *Suaeda salsa* and 1.4 times that of reeds [104].The decomposition of aboveground litter and the secretion of metabolites and residues from its well-developed underground roots into wetland soils directly affect the soil organic carbon content [105]. On the other hand, the tall stems and leaves and well-developed roots of *Spartina alterniflora* can weaken tidal dynamics, consolidate sediments on mudflats, and result in a large deposition of fine-grained particles from seawater, increasing organic carbon in the soil [106].

Taking the salt marshes of Yancheng, China, as an example, we note that although the total area has decreased by 43.09 km^2 over the past four decades, the high carbon sequestration capacity of *Spartina alterniflora* has strengthened the total carbon storage and burial in Yancheng's salt marshes [107]. These findings indicate that the carbon sequestration function of coastal blue carbon is largely regulated by the coverage of salt marsh vegetation and that the invasion of *Spartina alterniflora* to the seaward side of mudflats is beneficial for the carbon sequestration of salt marshes. Therefore, formulating environmental management measures to balance the negative ecological impacts of *Spartina alterniflora* and its strong carbon sequestration capacity is a complex challenge that requires more practical and systematic assessments to comprehensively consider its ecological effects.

4.3. Hydrological Regulation

Hydrological regulation is a common method used in wetland ecological restoration. For example, Hangzhou Bay National Wetland Park has fully leveraged the ecological functions of wetlands, such as maintaining biodiversity and improving water quality, through the restoration of hydrology [108]. The park has recorded 303 species of birds, 281 species of vascular plants, 192 species of phytoplankton and 121 species of large intertidal benthic animals, with an increasingly rich array of biological communities; the air quality within the park has reached the first-class national standard, and the water quality has improved significantly. Moreover, this restoration method is crucial for determining the impact on a wetland's carbon sequestration function. By adjusting the water levels in wetlands and increasing hydrological connectivity, the degraded hydrological environment can be effectively restored, which is vital for the growth and development of aquatic plants [109]. Aquatic plants absorb atmospheric carbon dioxide through photosynthesis, converting it into organic carbon stored within plants and soil, thereby increasing plant carbon reserves [96]. Moreover, the roots and litter of these plants can increase the amount of soil organic matter, slow the decomposition of organic matter, and increase the carbon storage capacity of wetlands.

Hydrological regulation not only directly affects the photosynthesis and respiration of plants but also indirectly influences the processes of wetland ecological functions. Suitable hydrological conditions can promote the growth of wetland vegetation and form different vegetation community structures, which have a decisive impact on the carbon source/sink function of wetland ecosystems [110]. For example, aquatic plants in wetlands, such as reeds and cattails, can form dense vegetation communities under suitable hydrological conditions. These communities not only absorb and store a large amount of carbon but also increase the organic carbon content in the soil through the decomposition of roots and litter [111]. In addition, hydrological regulation involves multiple aspects of wetlands, including water depth, water flow velocity and nutrient content in the water body. An appropriate water depth can provide a suitable growth environment for aquatic plants, and controlling the water flow velocity can reduce sediment erosion, protecting the organic carbon in wetland soils from being washed away [112]. Moreover, the nutrient content of a water body is crucial for the growth of aquatic plants. A suitable amount of nutrients can promote plant growth, whereas an excess may lead to water quality deterioration, affecting the health and carbon sequestration function of wetland ecosystems [113].

Hydrological regulation also involves the drainage and irrigation systems of wetlands. The rational design and management of these systems are equally important for maintaining the hydrological balance of wetlands and enhancing their carbon sequestration functions. Through scientifically sound and reasonable hydrological regulation, it is possible to effectively protect and restore wetland ecosystems, improve their adaptability and mitigation capacity to climate change, and thus play a significant role in the global carbon cycle and climate change.

4.4. Conversion of Farmland/Culture Ponds Back to Wetland

The implementation of protection and restoration projects such as converting farmland back to wetland, converting farmland to wetland, converting aquaculture to wetland, and converting fish ponds to wetland can help protect and restore wetland ecosystems and enhance carbon sequestration capabilities, which have a significant effect on enhancing the carbon sink function of wetlands [114]. These measures help reduce wetland degradation, restore the natural state of wetlands, and increase wetland area, thereby improving the carbon storage capacity of wetlands, and are crucial for enhancing the carbon sink functions of peatlands, coastal salt marshes, mangroves and other wetland ecosystems. They are essential for mitigating global climate change and contribute to the achievement of "dual carbon" goals through "nature-based climate solutions" [115]. For example, converting farmland back to wetland can restore wetland vegetation, increase the accumulation of soil organic carbon, and reduce the negative impact of agricultural activities on wetland carbon sinks. Additionally, converting aquaculture to wetlands can reduce the disturbance caused by fishing activities on wetland ecosystems, protect the natural ecological functions of wetlands and further enhance the carbon sink function of wetlands [116].

When a series of ecological restoration means, such as returning farmland to wetland, are carried out, the land use type is changed from agricultural land to wetland, which is actually equivalent to changing the type of wetland, and such a change will have an impact on soil carbon density (Figure 3). For example, after the land originally used for agricultural cultivation is returned to wetland, the soil structure and vegetation cover are changed, and the organic carbon content in the soil will gradually increase. With the gradual restoration of the wetland ecosystem, the soil carbon density may approach or even reach the carbon density level of the natural wetland shown in the figure, which will help enhance the carbon sink function of the wetland and thus play a positive role in mitigating climate change. Meanwhile, the soil carbon density varies among different types of wetlands; for example, the organic carbon density of farmed seawall soil is only 6.6 ($t \cdot hm^{-2}$), which is much lower than the average soil carbon density of natural wetlands.

Data analysis from some "pond-to-wetland" restoration projects in China revealed that the impact of "pond-to-wetland" ecological restoration projects on the carbon sequestration function of wetlands is significant. For example, the wetlands in the Liaohe Estuary "pond-to-wetland" ecological restoration area exhibited carbon sink characteristics in 2021, with a total net carbon sequestration rate of $-66.89\ g\cdot(m^2\cdot a)^{-1}$. Although this value is not as high as the carbon sequestration capacity of natural wetlands in the same area ($-161.92\ g\cdot(m^2\cdot a)^{-1}$), it is close to the carbon sequestration capacity of the Liaohe Estuary wetlands in 2005 ($-65.00\ g\cdot(m^2\cdot a)^{-1}$). These findings indicate that the "pond-to-wetland" ecological restoration project effectively restored the carbon sequestration function of the wetlands. Despite still producing a lower capacity than natural wetlands, the long-term potential of this project for increasing carbon sequestration should not be overlooked [117].

Furthermore, compared with other coastal wetlands, although the carbon sequestration capacity of the Liaohe Estuary "pond-to-wetland" ecological restoration area is not the most prominent, it has shown a positive trend of recovery. For example, compared with the annual carbon sequestration rates of natural wetlands in the Yangtze River Estuary ($-869.00\ g\cdot(m^2\cdot a)^{-1}$) and reed wetlands in the Yellow River Delta ($-599.05\ g\cdot(m^2\cdot a)^{-1}$), the carbon sequestration capacity of the Liaohe Estuary restoration area is lower, which may be related to the vegetation type, biomass and climatic conditions of the restoration area. However, the daily average carbon sequestration rate of the Liaohe Estuary "pond-to-wetland" ecological restoration area ($-0.55\ g\cdot(m^2\cdot a)^{-1}$) is higher than that of the salt marsh alkali wormwood wetlands in the Yellow River Estuary (-0.17~$-0.10\ g\cdot(m^2\cdot a)^{-1}$), which further shows the effectiveness of the restoration measures in enhancing the carbon sequestration function of wetlands.

In summary, "pond-to-wetland" ecological restoration projects have played a positive role in restoring and enhancing the carbon sequestration function of wetlands. These projects contribute not only to the recovery of wetland ecosystems but also to the global carbon cycle and the mitigation of climate change. With the deepening of restoration projects and the further recovery of wetland ecology, the carbon sequestration function of these areas is expected to be further strengthened.

4.5. Comparative Analysis of Restoration Project Types

By integrating all four restoration strategies analyzed in this study—vegetation plantation, invasive plant management, hydrological regulation and farmland/culture pond conversion—distinct advantages and limitations can be identified for each. Table 1 summarizes their respective contributions to carbon sink functions.

Table 1. Comparative analysis of carbon sink contributions from different wetland restoration strategies.

Restoration Type	Carbon Density Increase (%)	Carbon Fixation Rate (kg/m^2/year)	Long-Term Stability	Data Source
Vegetation Plantation	+30%	1.5	High	[18,79,90]
Invasive Plant Management	+20% (after recovery)	1.0	Moderate	[37,76,102]
Hydrological Regulation	+25%	1.2	High	[74,103]
Farmland/Culture Pond Conversion	+27%	1.3	High	[18,79,102]

Among the four strategies, vegetation plantation remains the most effective in rapidly enhancing carbon sink functions, primarily due to its direct contributions to biomass and soil organic carbon inputs. Hydrological regulation, while slightly less impactful in the short term, plays a crucial role in maintaining long-term stability by optimizing wetland hydrology and reducing methane emissions. The conversion of farmland or aquaculture ponds to wetlands offers significant carbon sequestration potential but requires careful consideration of socio-economic impacts and land use trade-offs.

Invasive plant management, although essential for biodiversity conservation, has slower carbon recovery rates due to the initial disturbance caused by plant removal. A potential solution is to integrate invasive species removal with vegetation plantation and hydrological regulation to maximize carbon storage and ecosystem restoration [79,102].

5. Implications of Incorporating Carbon Sink Enhancement into Wetland Ecological Restoration

In the process of wetland ecological restoration, it is vital to consider the enhancement of carbon sink function. This will not only help enhance the ecological service value of the wetlands themselves but also play a positive role in the global response to climate change. Wetland ecological restoration involves many aspects, such as plant cultivation, invasive plant management, hydrological regulation and land use type conversion, and each aspect is closely related to the carbon sink function.

5.1. Optimizing Plant Selection and Planting Strategies

In terms of plant cultivation, the selection of different plant species has a significant effect on wetland carbon sink functions [118]. For example, certain plants have special photosynthetic mechanisms or more efficient nutrient uptake capabilities that allow them to absorb more carbon dioxide under the same conditions and convert it to organic carbon storage [107,119]. Studies have shown that the biomass and carbon density of plants are closely related to their carbon fixation capacity [120]. Therefore, in restoration projects, plants with high biomass and high carbon density, such as reeds, should be preferentially selected for planting to improve the carbon storage of wetlands. Moreover, increasing the diversity of wetland plants also contributes to the stability of the ecosystem and the carbon sink function [121]. When multiple plants grow together, they complement each other in space and resource use, enabling ecosystems to respond more effectively to environmental changes such as climate change and water level fluctuations [122]. The differences in the root structure, growth cycle and ecological niche of different plants help increase soil porosity, improve soil structure and promote microbial activities, thereby improving the storage capacity of soil organic carbon and thus enhancing the ability of wetland ecosystems to absorb and store carbon.

5.2. Scientific Management of Invasive Plants

Invasive plant management is also an important part of wetland ecological restoration. Taking Spartina alterniflora as an example, we note that its impact on the carbon sink function of coastal wetlands is relatively complex [123]. Spartina alterniflora has strong reproduction and diffusion ability, strong growth, a long growing season, a high leaf area index and a high net photosynthetic rate, so it has high primary productivity and biomass [104]. Its expansion on mudflats and its own efficient carbon sequestration capacity have increased blue carbon storage in salt marshes, such as the salt marshes in Yancheng, China, where the high carbon sequestration capacity has increased the total carbon storage and embedment of salt marshes, although their total area has decreased by 43.09 km^2 over the past four decades [107]. However, the invasion of Spartina alterniflora has also changed the competition between native plants and resulted in certain disturbances to wetland ecosystems. Therefore, more practical and systematic assessments are needed to weigh the negative ecological impact and carbon sink capacity and formulate scientific management measures to achieve a balance between ecological protection and carbon sink function enhancement [124].

5.3. Rational Control of Hydrological Conditions

Hydrological regulation also plays a key role in enhancing the carbon sink function through wetland ecological restoration projects [125]. By adjusting the wetland water level and increasing hydrological connectivity, the degraded hydrological environment can be restored, which is very important for wetland carbon sink function [126]. Suitable hydrological conditions can promote the growth of aquatic plants such as reeds and cattails, which can absorb carbon dioxide through photosynthesis and convert it into organic carbon under appropriate water depth, water flow speed and water nutrition conditions [127]. Their roots and litter can also increase soil organic matter, slow decomposition and enhance the carbon storage capacity of wetlands [128]. Hydrological regulation involves various wetland factors [108]. The rational design and management of the drainage and irrigation systems of wetlands should maintain the balance of water depth, flow rate and nutrient content [129], which can not only provide a suitable growth environment for aquatic plants, reduce sediment erosion and protect soil organic carbon [130] but also prevent water quality deterioration caused by excess nutrients, thus maintaining the health of wetland ecosystems and improving their carbon sink function. The enhancement of climate change adaptation and mitigation capacity plays an important role in the global carbon cycle.

5.4. Promotion of Restoration Projects Such as Returning Farmlands to Wetlands

The implementation of protection and restoration projects such as returning farmlands to wetlands is highly important for improving the carbon sink function of wetlands. These projects help reduce wetland degradation, restore wetlands to their natural state and increase wetland area, thereby improving the carbon storage capacity of wetlands [84]. For example, restoring farmland to wetlands can restore wetland vegetation, increase soil organic carbon accumulation and reduce the negative impact of agricultural activities on wetland carbon sinks [131]. The conversion of aquaculture into wetlands can reduce the interference of fishery activities on wetland ecosystems, protect the natural ecological function of wetlands and further enhance the carbon sink function. Taking the Liaohe River Estuary "fish-pond-to-wetland" ecological restoration project as an example, we note that although its carbon sink capacity is still far from that of natural wetlands [56], it has shown a positive recovery trend and has the potential to increase carbon sinks in the long run, making significant contributions to wetland ecosystem restoration and the global carbon cycle.

5.5. General Discussion on Challenges and Future Directions

Despite the promising outcomes of these restoration strategies, several challenges remain: Site-Specific Variability: The effectiveness of restoration measures varies significantly depending on local environmental conditions such as salinity, sedimentation and nutrient availability. Tailored approaches are required to maximize their effectiveness in different regions [72,108]. Short-term vs. Long-term Trade-offs: Strategies like invasive plant management often result in short-term carbon losses, which may offset initial gains. Long-term monitoring is necessary to fully assess the recovery trends [107]. Economic and Social Impacts: Conversion of farmland or aquaculture ponds to wetlands may conflict with local economic activities, requiring careful planning and stakeholder engagement [84].

To address these challenges, future research should focus on the following: Integrated Restoration Approaches: Combining multiple strategies, such as vegetation plantation and hydrological regulation, to optimize both short-term and long-term carbon sequestration. Enhanced Monitoring Programs: Establishing long-term monitoring systems to track carbon fluxes and ecosystem changes across restored wetlands. Stakeholder Engagement: Involving local communities in planning and implementation to ensure socio-economic sustainability and project acceptance [108].

6. Conclusions

This study systematically analyzed the impact of various ecological restoration projects on the carbon sink function of coastal wetlands, clarifying the specific effects of projects such as vegetation planting, invasive species management, hydrological regulation and converting farmland to wetlands. The main contributions of this study include clarifying the significant enhancement of the carbon sink function through vegetation planting projects, particularly in terms of vegetation carbon storage and soil organic carbon content. It also revealed the complex impact of invasive species management on the carbon sink function, noting that while invasive species like Spartina alterniflora have high carbon fixation capabilities, their negative ecological impacts need to be comprehensively assessed. Additionally, the study elaborated on the role of hydrological regulation in enhancing the carbon sink function, emphasizing the importance of appropriate hydrological conditions for wetland vegetation growth and carbon sequestration. By comparing the carbon sink capacity before and after restoration measures such as converting farmland to wetlands, this study provided a scientific basis for planning and implementing ecological restoration projects, especially in enhancing carbon sink functions. Despite covering various types of ecological restoration projects, this study has limitations, such as short monitoring periods for some projects that may not fully reflect long-term impacts on the carbon sink function. Moreover, data collection for restoration projects in specific ecosystem types, such as mangrove restoration, was insufficient. Future research should extend monitoring periods, expand the scope of study and conduct comparative analyses across more ecosystem types to more comprehensively evaluate the impact of ecological restoration projects on the carbon sink function.

Author Contributions: Conceptualization, Q.W. and X.G.; methodology, Y.L. (Yanjin Liu); software, Y.L. (Yina Li); writing—original draft preparation, X.G.; writing—review and editing, Q.W., T.X. and H.L. All authors have read and agreed to the published version of the manuscript.

Funding: This study was supported financially by the Natural Science Foundation of China (42107057, U2243208, 52271256, 42330705) and the Science & Technology Fundamental Resources Investigation Program (Grant No. 2022FY100304).

Data Availability Statement: The raw data supporting the conclusions of this article will be made available by the authors upon request.

Conflicts of Interest: The authors declare that the research was conducted in the absence of any commercial or financial relationships that could be construed as a potential conflict of interest.

References

1. Lovelock, C.E.; Duarte, C.M. Dimensions of Blue Carbon and emerging perspectives. *Biol. Lett.* **2019**, *15*, 3. [CrossRef]
2. Jiao, N.Z.; Liu, J.H.; Shi, T.; Zhang, C.L.; Zhang, Y.Y.; Zheng, Q.; Chen, Q.R.; Tang, K.; Wang, Y.Z.; Dong, H.L. Implementing negative ocean emissions and implementing carbon neutral strategies. *Sci. China Earth Sci.* **2021**, *51*, 632–643. [CrossRef]
3. Dai, M.; Su, J.; Zhao, Y.; Hofmann, E.E.; Cao, Z.; Cai, W.-J.; Gan, J.; Lacroix, F.; Laruelle, G.G.; Meng, F.; et al. Carbon Fluxes in the Coastal Ocean: Synthesis, Boundary Processes, and Future Trends. *Annu. Rev. Earth Planet. Sci.* **2022**, *50*, 593–626. [CrossRef]
4. IPCC; Stocker, T.F.; Qin, D.; Plattner, G.K.; Midgley, P.M. *The Physical Science Basis. Contribution of Working Group into the Fifth Assessment Report of the Intergovernmental Panel on Climate Change*; Cambridge University Press: Cambridge, UK, 2013. [CrossRef]
5. Han, G.; Wang, F.; Ma, J.; Xiao, L.; Chu, X.; Zhao, M. Blue carbon sink function, formation mechanism and sequestration potential of coastal salt marshes. *Chin. J. Plant Ecol.* **2022**, *46*, 373–382. [CrossRef]
6. Alongi, D.M. Carbon sequestration in mangrove forests. *Carbon Manag.* **2012**, *3*, 313–322. [CrossRef]
7. Cadiz, P.L.; Calumpong, H.P.; Sinutok, S.; Chotikarn, P. Carbon storage potential of natural and planted mangals in trang, thailand. *Appl. Ecol. Environ. Res.* **2020**, *18*, 4383–4403. [CrossRef]
8. Donato, D.C.; Kauffman, J.B.; Murdiyarso, D.; Kurnianto, S.; Stidham, M.; Kanninen, M. Mangroves among the most carbon-rich forests in the tropics. *Nat. Geosci.* **2011**, *4*, 293–297. [CrossRef]

9. Barbier, E.B.; Hacker, S.D.; Kennedy, C.; Koch, E.W.; Stier, A.C.; Silliman, B.R. The value of estuarine and coastal ecosystem services. *Ecol. Monogr.* **2011**, *81*, 169–193. [CrossRef]
10. Leonardi, N.; Ganju, N.K.; Fagherazzi, S. A linear relationship between wave power and erosion determines salt-marsh resilience to violent storms and hurricanes. *Proc. Natl. Acad. Sci. USA* **2016**, *113*, 64–68. [CrossRef] [PubMed]
11. Perillo, G.; Wolanski, E.; Cahoon, D.R.; Hopkinson, C.S. *Coastal Wetlands: An Integrated Ecosystem Approach*; Elsevier: Amsterdam, The Netherlands, 2019. [CrossRef]
12. Ramus, A.P.; Silliman, B.R.; Thomsen, M.S.; Long, Z.T. An invasive foundation species enhances multifunctionality in a coastal ecosystem. *Proc. Natl. Acad. Sci. USA* **2017**, *114*, 8580–8585. [CrossRef]
13. Worm, B.; Barbier, E.B.; Beaumont, N.; Duffy, J.E.; Folke, C.; Halpern, B.S.; Jackson, J.B.C.; Lotze, H.K.; Micheli, F.; Palumbi, S.R.; et al. Impacts of biodiversity loss on ocean ecosystem services. *Science* **2006**, *314*, 787–790. [CrossRef] [PubMed]
14. Wang, X.; Xiao, X.; Xu, X.; Zou, Z.; Chen, B.; Qin, Y.; Zhang, X.; Dong, J.; Liu, D.; Pan, L.; et al. Rebound in China's coastal wetlands following conservation and restoration. *Nat. Sustain.* **2021**, *4*, 1076–1083. [CrossRef]
15. Osland, M.J.; Chivoiu, B.; Enwright, N.M.; Thorne, K.M.; Guntenspergen, G.R.; Grace, J.B.; Dale, L.L.; Brooks, W.; Herold, N.; Day, J.W.; et al. Migration and transformation of coastal wetlands in response to rising seas. *Sci. Adv.* **2022**, *8*, 26. [CrossRef] [PubMed]
16. Tang, J.; Ye, S.; Chen, X.; Yang, H.; Sun, X.; Wang, F.; Wen, Q.; Chen, S. Coastal blue carbon: Concept, study method, and the application to ecological restoration. *Sci. China Earth Sci.* **2018**, *61*, 637–646. [CrossRef]
17. Czapla, K.M.; Anderson, I.C.; Currin, C.A. Net Ecosystem Carbon Balance in a North Carolina, USA, Salt Marsh. *J. Geophys. Res.-Biogeosci.* **2020**, *125*, 10. [CrossRef]
18. Choudhary, B.; Dhar, V.; Pawase, A.S. Blue carbon and the role of mangroves in carbon sequestration: Its mechanisms, estimation, human impacts and conservation strategies for economic incentives. *J. Sea Res.* **2024**, *199*, 102504. [CrossRef]
19. Forbrich, I.; Giblin, A.E.; Hopkinson, C.S. Constraining Marsh Carbon Budgets Using Long-Term C Burial and Contemporary Atmospheric CO_2 Fluxes. *J. Geophys. Res. Biogeosci.* **2018**, *123*, 867–878. [CrossRef]
20. Chen, L.; Wang, W.; Zhang, Y.; Lin, G. Recent progresses in mangrove conservation, restoration and research in China. *J. Plant Ecol.* **2009**, *2*, 45–54. [CrossRef]
21. Wang, C.; Liu, H.; Zhu, L.; Ren, H.; Yan, J.; Li, Z.; Zhang, H. Which traits are necessary to quickly select suitable plant species for ecological restoration? *Ecol. Solut. Evid.* **2021**, *2*, 4. [CrossRef] [PubMed]
22. Chen, Y.; Chen, L. Interactions between vegetation and sediment carbon pools within coastal blue carbon ecosystems: A review and perspective. *J. Mar. Sci.* **2023**, *41*, 3–13. [CrossRef]
23. Han, G.; Sun, B.; Chu, X.; Xing, Q.; Song, W.; Xia, J. Precipitation events reduce soil respiration in a coastal wetland based on four-year continuous field measurements. *Agric. For. Meteorol.* **2018**, *256*, 292–303. [CrossRef]
24. de Lacerda, L.D.; Marins, R.V.; da Silva Dias, F.J. An Arctic Paradox: Response of Fluvial Hg Inputs and Bioavailability to Global Climate Change in an Extreme Coastal Environment. *Front. Earth Sci.* **2020**, *8*, 93. [CrossRef]
25. Newton, A.; Icely, J.; Cristina, S.; Perillo, G.M.E.; Turner, R.E.; Ashan, D.; Cragg, S.; Luo, Y.; Tu, C.; Li, Y.; et al. Anthropogenic, Direct Pressures on Coastal Wetlands. *Front. Ecol. Evol.* **2020**, *8*, 144. [CrossRef]
26. Cvetkovic, M.; Chow-Fraser, P. Use of ecological indicators to assess the quality of Great Lakes coastal wetlands. *Ecol. Indic.* **2011**, *11*, 1609–1622. [CrossRef]
27. Newton, A.; Carruthers, T.J.B.; Icely, J. The coastal syndromes and hotspots on the coast. *Estuar. Coast. Shelf Sci.* **2012**, *96*, 39–47. [CrossRef]
28. Valiela, I.; Kinney, E.L.; Smith, S.; Culbertson, J. Global losses of mangroves and salt marshes. In: Duarte, C.M. (Ed.), Global Loss of Coastal Habitats Rates. *Causes Conseq.* **2009**, *1*, 109–142.
29. Murray, N.J.; Clemens, R.S.; Phinn, S.R.; Possingham, H.P.; Fuller, R.A. Tracking the rapid loss of tidal wetlands in the Yellow Sea. *Front. Ecol. Environ.* **2014**, *12*, 267–272. [CrossRef]
30. Lal, R.; Pimentel, D. Soil erosion: A carbon sink or source? *Science* **2008**, *319*, 1040–1042. [CrossRef] [PubMed]
31. Lane, R.R.; Mack, S.K.; Day, J.W.; DeLaune, R.D.; Madison, M.J.; Precht, P.R. Fate of Soil Organic Carbon During Wetland Loss. *Wetland* **2016**, *36*, 1167–1181. [CrossRef]
32. Rey Benayas, J.M.; Newton, A.C.; Diaz, A.; Bullock, J.M. Enhancement of Biodiversity and Ecosystem Services by Ecological Restoration: A Meta-Analysis. *Science* **2009**, *325*, 1121–1124. [CrossRef] [PubMed]
33. Wortley, L.; Hero, J.-M.; Howes, M. Evaluating Ecological Restoration Success: A Review of the Literature. *Restor. Ecol.* **2013**, *21*, 537–543. [CrossRef]
34. Zhou, D.; Yu, J.; Guan, B.; Li, Y.; Yu, M.; Qu, F.; Zhan, C.; Lv, Z.; Wu, H.; Wang, Q.; et al. A Comparison of the Development of Wetland Restoration Techniques in China and Other Nations. *Wetland* **2020**, *40*, 2755–2764. [CrossRef]
35. Rodrigo, M.A. Wetland Restoration with Hydrophytes: A Review. *Plants* **2021**, *10*, 1035. [CrossRef] [PubMed]
36. Na, T.; BaoShan, C.U.I.; XinSheng, Z. The restoration of reed(*Phragmites australis*) wetland in the Yellow River Delta. *Acta Ecol. Sin.* **2006**, *26*, 2616–2624.

37. Fan, J.; Wang, X.; Wu, W.; Chen, W.; Ma, Q.; Ma, Z. Function of restored wetlands for waterbird conservation in the Yellow Sea coast. *Sci. Total Environ.* **2021**, *756*, 144061. [CrossRef] [PubMed]
38. Pacala, S.W.; Hurtt, G.C.; Baker, D.; Peylin, P.; Houghton, R.A.; Birdsey, R.A.; Heath, L.; Sundquist, E.T.; Stallard, R.F.; Ciais, P.; et al. Consistent land- and atmosphere-based US carbon sink estimates. *Science* **2001**, *292*, 2316–2320. [CrossRef]
39. Xu, X.; Liu, H.; Liu, Y.; Zhou, C.; Pan, L.; Fang, C.; Nie, M.; Li, B. Human eutrophication drives biogeographic salt marsh productivity patterns in China. *Ecol. Appl.* **2020**, *30*, e02045. [CrossRef] [PubMed]
40. Choi, Y.H.; Wang, Y. Dynamics of carbon sequestration in a coastal wetland using radiocarbon measurements. *GBC* **2004**, *18*, GB4016. [CrossRef]
41. Lian, X.; Jeong, S.; Park, C.-E.; Xu, H.; Li, L.Z.X.; Wang, T.; Gentine, P.; Penuelas, J.; Piao, S. Biophysical impacts of northern vegetation changes on seasonal warming patterns. *Nat. Commun.* **2022**, *13*, 3925. [CrossRef] [PubMed]
42. Karlowsky, S.; Augusti, A.; Ingrisch, J.; Hasibeder, R.; Lange, M.; Lavorel, S.; Bahn, M.; Gleixner, G. Land use in mountain grasslands alters drought response and recovery of carbon allocation and plant-microbial interactions. *J. Ecol.* **2018**, *106*, 1230–1243. [CrossRef] [PubMed]
43. Diao, H.; Wang, A.; Yuan, F.; Guan, D.; Dai, G.; Wu, J. Environmental Effects on Carbon Isotope Discrimination from Assimilation to Respiration in a Coniferous and Broad-Leaved Mixed Forest of Northeast China. *Forests* **2020**, *11*, 1156. [CrossRef]
44. Williams, A.; de Vries, F.T. Plant root exudation under drought: Implications for ecosystem functioning. *New Phytol* **2020**, *225*, 1899–1905. [CrossRef]
45. Preece, C.; Farre-Armengol, G.; Llusia, J.; Penuelas, J. Thirsty tree roots exude more carbon. *Tree Physiol.* **2018**, *38*, 690–695. [CrossRef]
46. Pausch, J.; Kuzyakov, Y. Carbon input by roots into the soil: Quantification of rhizodeposition from root to ecosystem scale. *Glob. Change Biol.* **2018**, *24*, 1–12. [CrossRef] [PubMed]
47. Asmala, E.; Haraguchi, L.; Markager, S.; Massicotte, P.; Riemann, B.; Staehr, P.A.; Carstensen, J. Eutrophication Leads to Accumulation of Recalcitrant Autochthonous Organic Matter in Coastal Environment. *Glob. Biogeochem. Cycles* **2018**, *32*, 1673–1687. [CrossRef]
48. Xia, S.; Song, Z.; Van Zwieten, L.; Guo, L.; Yu, C.; Wang, W.; Li, Q.; Hartley, I.P.; Yang, Y.; Liu, H.; et al. Storage, patterns and influencing factors for soil organic carbon in coastal wetlands of China. *Glob. Change Biol.* **2022**, *28*, 6065–6085. [CrossRef] [PubMed]
49. Deb, S.; Mandal, B. Soils and sediments of coastal ecology: A global carbon sink. *Ocean Coast. Manag.* **2021**, *214*, 105937. [CrossRef]
50. Breithaupt, J.L.; Smoak, J.M.; Bianchi, T.S.; Vaughn, D.R.; Sanders, C.J.; Radabaugh, K.R.; Osland, M.J.; Feher, L.C.; Lynch, J.C.; Cahoon, D.R.; et al. Increasing Rates of Carbon Burial in Southwest Florida Coastal Wetlands. *J. Geophys. Res. Biogeosci.* **2020**, *125*, 5349. [CrossRef]
51. Ishtiaq, K.S.; Troxler, T.G.; Lamb-Wotton, L.; Wilson, B.J.; Charles, S.P.; Davis, S.E.; Kominoski, J.S.; Rudnick, D.T.; Sklar, F.H. Modeling net ecosystem carbon balance and loss in coastal wetlands exposed to sea-level rise and saltwater intrusion. *Ecol. Appl.* **2022**, *32*, 8. [CrossRef] [PubMed]
52. Jarvis, N.; Coucheney, E.; Lewan, E.; Kloeffel, T.; Meurer, K.H.E.; Keller, T.; Larsbo, M. Interactions between soil structure dynamics, hydrological processes, and organic matter cycling: A new soil-crop model. *Eur. J. Soil Sci.* **2024**, *75*, 2. [CrossRef]
53. Vedere, C.; Lebrun, M.; Honvault, N.; Aubertin, M.-L.; Girardin, C.; Garnier, P.; Dignac, M.-F.; Houben, D.; Rumpel, C. How does soil water status influence the fate of soil organic matter? A review of processes across scales. *Earth Sci. Rev.* **2022**, *234*, 104214. [CrossRef]
54. Li, J.-Y.; Chen, P.; Li, Z.-G.; Li, L.-Y.; Zhang, R.-Q.; Hu, W.; Liu, Y. Soil aggregate-associated organic carbon mineralization and its driving factors in rhizosphere soil. *SBB* **2023**, *186*, 109182. [CrossRef]
55. Adeleke, R.; Nwangburuka, C.; Oboirien, B. Origins, roles and fate of organic acids in soils: A review. *S. Afr. J. Bot.* **2017**, *108*, 393–406. [CrossRef]
56. Villa, J.A.; Bernal, B. Carbon sequestration in wetlands, from science to practice: An overview of the biogeochemical process, measurement methods, and policy framework. *Ecol. Eng.* **2018**, *114*, 115–128. [CrossRef]
57. Morrissey, E.M.; Kane, J.; Tripathi, B.M.; Rion, M.S.I.; Hungate, B.A.; Franklin, R.; Walter, C.; Sulman, B.; Brzostek, E. Carbon acquisition ecological strategies to connect soil microbial biodiversity and carbon cycling. *SBB* **2023**, *177*, 108893. [CrossRef]
58. Wang, J.; Yu, G.; Han, L.; Yao, Y.; Sun, M.; Yan, Z. Ecosystem carbon exchange across China's coastal wetlands: Spatial patterns, mechanisms, and magnitudes. *Agric. For. Meteorol.* **2024**, *345*, 109859. [CrossRef]
59. Li, X.; Cheng, X.; Cheng, K.; Cai, Z.; Feng, S.; Zhou, J. The influence of tide-brought nutrients on microbial carbon metabolic profiles of mangrove sediments. *Sci. Total Environ.* **2024**, *906*, 167732. [CrossRef]
60. Li, S.; Li, Y.; Jiang, N.; Xu, W. Development of key ecological conservation and restoration projects in the past century. *Ecol. Front.* **2025**, *45*, 1–6. [CrossRef]

61. Day, J.W.; Twilley, R.R.; Freeman, A.; Couvillion, B.; Quirk, T.; Jafari, N.; Mariotti, G.; Hunter, R.; Norman, C.; Kemp, G.P.; et al. The concept of land bridge marshes in the Mississippi River Delta and implications for coastal restoration. *Nat.-Based Solut.* **2023**, *3*, 100061. [CrossRef]
62. Sklar, F.H.; Meeder, J.F.; Troxler, T.G.; Dreschel, T.; Davis, S.E.; Ruiz, P.L. Chapter 16—The Everglades: At the Forefront of Transition. In *Coasts Estuaries*; Elsevier: Amsterdam, The Netherlands, 2019; pp. 277–292. [CrossRef]
63. Ma, X.; Brookes, J.; Wang, X.; Han, Y.; Ma, J.; Li, G.; Chen, Q.; Zhou, S.; Qin, B. Water quality improvement and existing challenges in the Pearl River Basin, China. *J. Water Process. Eng.* **2023**, *55*, 104184. [CrossRef]
64. Cheng, C.; Li, F. Ecosystem restoration and management based on nature-based solutions in China: Research progress and representative practices. *Nat. Based Sol.* **2024**, *6*, 100176. [CrossRef]
65. Jiang, Y.; Xiao, Y.; Zhang, Z.; Zhao, S. How does central-local interaction affect local environmental governance? Insights from the transformation of central environmental protection inspection in China. *Environ. Res.* **2024**, *243*, 117668. [CrossRef] [PubMed]
66. Cai, Y.; Liang, J.; Zhang, P.; Wang, Q.; Wu, Y.; Ding, Y.; Wang, H.; Fu, C.; Sun, J. Review on strategies of close-to-natural wetland restoration and a brief case plan for a typical wetland in northern China. *Chemosphere* **2021**, *285*, 131534. [CrossRef] [PubMed]
67. Sivapalan, M.; Bloeschl, G. The Growth of Hydrological Understanding: Technologies, Ideas, and Societal Needs Shape the Field. *Water Resour. Res.* **2017**, *53*, 8137–8146. [CrossRef]
68. Du, J.; Chen, B.; Nagelkerken, I.; Chen, S.; Hu, W. Protect seagrass meadows in China's waters. *Science* **2023**, *379*, 447. [CrossRef] [PubMed]
69. Zhang, X.; Xiao, X.; Wang, X.; Xu, X.; Qiu, S.; Pan, L.; Ma, J.; Ju, R.; Wu, J.; Li, B. Continual expansion of *Spartina alterniflora* in the temperate and subtropical coastal zones of China during 1985–2020. *Int. J. Appl. Earth Obs. Geoinf.* **2023**, *117*, 103192. [CrossRef]
70. Sun, Z.; Sun, W.; Tong, C.; Zeng, C.; Yu, X.; Mou, X. China's coastal wetlands: Conservation history, implementation efforts, existing issues and strategies for future improvement. *Environ. Int.* **2015**, *79*, 25–41. [CrossRef] [PubMed]
71. Tang, C.; Liu, Y.; Li, Z.; Guo, L.; Xu, A.; Zhao, J. Effectiveness of vegetation cover pattern on regulating soil erosion and runoff generation in red soil environment, southern China. *Ecol. Indic.* **2021**, *129*, 107956. [CrossRef]
72. Yang, R.; Yang, S.; Chen, L.-l.; Yang, Z.; Xu, L.; Zhang, X.; Liu, G.; Jiao, C.; Bai, R.; Zhang, X.; et al. Effect of vegetation restoration on soil erosion control and soil carbon and nitrogen dynamics: A meta-analysis. *Soil Till Res.* **2023**, *230*, 105705. [CrossRef]
73. Thomas, E.; Jalonen, R.; Loo, J.; Boshier, D.; Gallo, L.; Cavers, S.; Bordacs, S.; Smith, P.; Bozzano, M. Genetic considerations in ecosystem restoration using native tree species. *For. Ecol. Manag.* **2014**, *333*, 66–75. [CrossRef]
74. He, M.; Cui, J.; Yi, Y.; Chen, H.W.; Zhang, Q.; Li, L.; Huang, L.; Hong, S. Vegetation increases global climate vulnerability risk by shifting climate zones in response to rising atmospheric CO_2. *Sci. Total Environ.* **2024**, *949*, 174810. [CrossRef] [PubMed]
75. Aradottir, A.L.; Hagen, D. Chapter Three—Ecological Restoration: Approaches and Impacts on Vegetation, Soils and Society. *Adv. Agron.* **2013**, *120*, 173–222. [CrossRef]
76. Dou, Y.; Tong, X.; Horion, S.; Feng, L.; Fensholt, R.; Shao, Q.; Tian, F. The success of ecological engineering projects on vegetation restoration in China strongly depends on climatic conditions. *Sci. Total Environ.* **2024**, *915*, 170041. [CrossRef] [PubMed]
77. Zhu, Q.; Cai, Y. Integrating ecological risk, ecosystem health, and ecosystem services for assessing regional ecological security and its driving factors: Insights from a large river basin in China. *Ecol. Indic.* **2023**, *155*, 110954. [CrossRef]
78. Maity, R.; Srivastava, A.; Sarkar, S.; Khan, M.I. Revolutionizing the future of hydrological science: Impact of machine learning and deep learning amidst emerging explainable AI and transfer learning. *Appl. Comput. Geosci.* **2024**, *24*, 100206. [CrossRef]
79. Lorrain-Soligon, L.; Robin, F.; Bertin, X.; Jankovic, M.; Rousseau, P.; Lelong, V.; Brischoux, F. Long-term trends of salinity in coastal wetlands: Effects of climate, extreme weather events, and sea water level. *Environ. Res.* **2023**, *237*, 116937. [CrossRef]
80. Sun, J.; Chen, W.; Hu, B.; Xu, Y.J.; Zhang, G.; Wu, Y.; Hu, B.; Song, Z. Roles of reservoirs in regulating basin flood and droughts risks under climate change: Historical assessment and future projection. *J. Hydrol. Reg. Stud.* **2023**, *48*, 101453. [CrossRef]
81. Duan, H.; Xu, M.; Cai, Y.; Wang, X.; Zhou, J.; Zhang, Q. A Holistic Wetland Ecological Water Replenishment Scheme with Consideration of Seasonal Effect. *Sustainability* **2019**, *11*, 930. [CrossRef]
82. Li, Y.; Gu, X.; Zhou, B.; Tian, J.; Liu, H. Evaluation on the ecological reconstruction of aquatic communities in the coastal constructed wetland of northern China: A case study of Tianjin Lingang constructed wetland (phase II). *CJEE* **2022**, *16*, 1713–1720. [CrossRef]
83. Salimi, S.; Almuktar, S.A.A.A.N.; Scholz, M. Impact of climate change on wetland ecosystems: A critical review of experimental wetlands. *J. Environ. Manag.* **2021**, *286*, 112160. [CrossRef]
84. Endter-Wada, J.; Kettenring, K.M.; Sutton-Grier, A. Protecting wetlands for people: Strategic policy action can help wetlands mitigate risks and enhance resilience. *Environ. Sci. Policy* **2020**, *108*, 37–44. [CrossRef]
85. Schuster, L.; Taillardat, P.; Macreadie, P.I.; Malerba, M.E. Freshwater wetland restoration and conservation are long-term natural climate solutions. *Sci. Total Environ.* **2024**, *922*, 171218. [CrossRef]
86. Zhang, X.; Chen, P.; Han, Y.; Chang, X.; Dai, S. Quantitative analysis of self-purification capacity of non-point source pollutants in watersheds based on SWAT model. *Ecol. Indic.* **2022**, *143*, 109425. [CrossRef]

87. Ferreira, C.S.S.; Kasanin-Grubin, M.; Solomun, M.K.; Sushkova, S.; Minkina, T.; Zhao, W.; Kalantari, Z. Wetlands as nature-based solutions for water management in different environments. *Curr. Opin. Environ. Sci. Health* **2023**, *33*, 100476. [CrossRef]
88. Liu, N.; Ma, Z. Ecological restoration of coastal wetlands in China: Current status and suggestions. *Biol. Conserv.* **2024**, *291*, 110513. [CrossRef]
89. Ma, Z.; Lu, M.; Jin, H.; Sheng, X.; Wei, H.; Yang, Q.; Qi, L.; Huang, J.; Chen, L.; Dou, X. Greenhouse gas emissions and environmental drivers in different natural wetland regions of China. *Environ. Pollut.* **2023**, *330*, 121754. [CrossRef] [PubMed]
90. Chen, K.; Li, T.; Yang, M.; Zhou, X.; Peng, C. The effects of environmental factors and plant diversity on forest carbon sequestration vary between eastern and western regions of China. *J. Clean. Prod.* **2024**, *437*, 140371. [CrossRef]
91. Qiu, Z.; Feng, Z.; Song, Y.; Li, M.; Zhang, P. Carbon sequestration potential of forest vegetation in China from 2003 to 2050: Predicting forest vegetation growth based on climate and the environment. *J. Clean. Prod.* **2020**, *252*, 119715. [CrossRef]
92. Zhu, M.; Zhou, Z.; Wu, X.; Liu, R.; Zheng, J.; Wang, J.; Wan, J. Response of vegetation carbon sequestration potential to the effectiveness of vegetation restoration in karst ecologically fragile areas in Guizhou, southwest China. *Ecol. Indic.* **2024**, *158*, 111495. [CrossRef]
93. Semchenko, M.; Xue, P.; Leigh, T. Functional diversity and identity of plant genotypes regulate rhizodeposition and soil microbial activity. *New Phytol* **2021**, *232*, 776–787. [CrossRef] [PubMed]
94. Wang, K.; She, D.; Zhang, X.; Wang, Y.; Wen, H.; Yu, J.; Wang, Q.; Han, S.; Wang, W. Tree richness increased biomass carbon sequestration and ecosystem stability of temperate forests in China: Interacted factors and implications. *J. Environ. Manag.* **2024**, *368*, 122214. [CrossRef] [PubMed]
95. Cao, L.; Song, J.; Li, X.; Yuan, H.; Li, N.; Duan, L. Research progresses in carbon budget and carbon cycle of the coastal salt marshes in China. *Acta Ecol. Sin.* **2013**, *33*, 5141–5152. [CrossRef]
96. Li, B.; Gao, G.; Luo, Y.; Xu, M.; Liu, G.; Fu, B. Carbon stock and sequestration of planted and natural forests along climate gradient in water-limited area: A synthesis in the China's Loess plateau. *Agric. For. Meteorol.* **2023**, *333*, 109419. [CrossRef]
97. Dusenge, M.E.; Duarte, A.G.; Way, D.A. Plant carbon metabolism and climate change: Elevated CO_2 and temperature impacts on photosynthesis, photorespiration and respiration. *New Phytol.* **2019**, *221*, 32–49. [CrossRef] [PubMed]
98. Zhou, Z.; Lu, J.-Z.; Widyastuti, R.; Scheu, S.; Potapov, A.; Krashevska, V. Plant roots are more strongly linked to microorganisms in leaf litter rather than in soil across tropical land-use systems. *SBB* **2024**, *190*, 109320. [CrossRef]
99. Token, S.; Jiang, L.; Zhang, L.; Ly, G. Effects of plant diversity on primary productivity and community stability along soil water and salinity gradients. *Glob. Ecol. Conserv.* **2022**, *36*, e02095. [CrossRef]
100. Bluethgen, N.; Klein, A.-M. Functional complementarity and specialisation: The role of biodiversity in plant-pollinator interactions. *Basic Appl. Ecol.* **2011**, *12*, 282–291. [CrossRef]
101. Amyntas, A.; Berti, E.; Gauzens, B.; Albert, G.; Yu, W.; Werner, A.S.; Eisenhauer, N.; Brose, U. Niche complementarity among plants and animals can alter the biodiversity-ecosystem functioning relationship. *Funct. Ecol.* **2023**, *37*, 2652–2665. [CrossRef]
102. Zhao, Z.-y.; Xu, Y.; Yuan, L.; Li, W.; Zhu, X.-j.; Zhang, L.-q. Emergency control of *Spartina alterniflora* re-invasion with a chemical method in Chongming Dongtan, China. *WSE* **2020**, *13*, 24–33. [CrossRef]
103. Sun, L.; Shao, D.; Xie, T.; Gao, W.; Ma, X.; Ning, Z.; Cui, B. How Does *Spartina alterniflora* Invade in Salt Marsh in Relation to Tidal Channel Networks? Patterns and Processes. *Remote Sens.* **2020**, *12*, 2983. [CrossRef]
104. Zhou, C.; Zhao, H.; Sun, Z.; Zhou, L.; Fang, C.; Xiao, Y.; Deng, Z.; Zhi, Y.; Zhao, Y.; An, S. The Invasion of *Spartina alterniflora* Alters Carbon Dynamics in China's Yancheng Natural Reserve. *Clean-Soil Air Water* **2015**, *43*, 159–165. [CrossRef]
105. Jiang, L.-F.; Luo, Y.-Q.; Chen, J.-K.; Li, B. Ecophysiological characteristics of invasive *Spartina alterniflora* and native species in salt marshes of Yangtze River estuary, China. *Estuar. Coast. Shelf Sci.* **2009**, *81*, 74–82. [CrossRef]
106. Raza, T.; Qadir, M.F.; Khan, K.S.; Eash, N.S.; Yousuf, M.; Chatterjee, S.; Manzoor, R.; Rehman, S.u.; Oetting, J.N. Unraveling the potential of microbes in decomposition of organic matter and release of carbon in the ecosystem. *J. Environ. Manag.* **2023**, *344*, 118529. [CrossRef] [PubMed]
107. Wan, S.; Qin, P.; Liu, J.; Zhou, H. The positive and negative effects of exotic *Spartina alterniflora* in China. *Ecol. Eng.* **2009**, *35*, 444–452. [CrossRef]
108. Zhou, X.; Hu, C.; Wang, Z. Distribution of biomass and carbon content in estimation of carbon density for typical forests. *Glob. Ecol. Conserv.* **2023**, *48*, e02707. [CrossRef]
109. Zeng, X.; Zhang, H.; Zhou, B.; Liang, X.; Cui, L.; Li, H.; Qu, Y.; Luo, C. Hydrological dynamics and its impact on wetland ecological functions in the Sanjiang Plain, China. *Ecol. Indic.* **2024**, *169*, 112878. [CrossRef]
110. Twomey, A.J.; Nunez, K.; Carr, J.A.; Crooks, S.; Friess, D.A.; Glamore, W.; Orr, M.; Reef, R.; Rogers, K.; Waltham, N.J.; et al. Planning hydrological restoration of coastal wetlands: Key model considerations and solutions. *Sci. Total Environ.* **2024**, *915*, 169881. [CrossRef] [PubMed]
111. Wei, D.; Zhang, X.; Wang, X. Strengthening Hydrological Regulation of China's Wetland Greenness Under a Warmer Climate. *J. Geophys. Res. Biogeosci.* **2017**, *122*, 3206–3217. [CrossRef]

112. Pan, Y.; Cieraad, E.; Clarkson, B.R.; Colmer, T.D.; Pedersen, O.; Visser, E.J.W.; Voesenek, L.A.C.J.; van Bodegom, P.M. Drivers of plant traits that allow survival in wetlands. *Funct. Ecol.* **2020**, *34*, 956–967. [CrossRef]
113. O'Hare, M.T.; Aguiar, F.C.; Asaeda, T.; Bakker, E.S.; Chambers, P.A.; Clayton, J.S.; Elger, A.; Ferreira, T.M.; Gross, E.M.; Gunn, I.D.M.; et al. Plants in aquatic ecosystems: Current trends and future directions. *Hydrobiologia* **2018**, *812*, 1–11. [CrossRef]
114. Lambert, S.J.; Davy, A.J. Water quality as a threat to aquatic plants: Discriminating between the effects of nitrate, phosphate, boron and heavy metals on charophytes. *New Phytol* **2011**, *189*, 1051–1059. [CrossRef] [PubMed]
115. Xu, X.-C.; Zhao, X.-Y.; Song, X.-Y. Impacts of the returning farmland to forest (grassland)project on ecosystem services in the Weihe River Basin, China. *J. Appl. Ecol.* **2021**, *32*, 3893–3904. [CrossRef]
116. Fan, B.; Li, Y. China's conservation and restoration of coastal wetlands offset much of the reclamation-induced blue carbon losses. *Glob. Change Biol.* **2024**, *30*, e17039. [CrossRef] [PubMed]
117. Zhang, G.S.; Cai, Y.Y.; Yan, J.S.; Ma, Y.; Gong, W.; Wang, C.J. Impact of reclamation on net ecosystem CO_2 exchange in the Liaohe Estuary salt marsh wetlands. *Acta Phytoecol. Sin.* **2021**, *43*, 18–22. [CrossRef]
118. Lemoine, T.; Violle, C.; Montazeaud, G.; Isaac, M.E.; Rocher, A.; Freville, H.; Fort, F. Plant trait relationships are maintained within a major crop species: Lack of artificial selection signal and potential for improved agronomic performance. *New Phytol.* **2023**, *240*, 2227–2238. [CrossRef] [PubMed]
119. Tariq, A.; Graciano, C.; Sardans, J.; Zeng, F.; Hughes, A.C.; Ahmed, Z.; Ullah, A.; Ali, S.; Gao, Y.; Penuelas, J. Plant root mechanisms and their effects on carbon and nutrient accumulation in desert ecosystems under changes in land use and climate. *New Phytol.* **2024**, *242*, 916–934. [CrossRef] [PubMed]
120. Jin, Y.; Liu, C.; Qian, S.S.; Luo, Y.; Zhou, R.; Tang, J.; Bao, W. Large-scale patterns of understory biomass and its allocation across China's forests. *Sci. Total Environ.* **2022**, *804*, 150169. [CrossRef]
121. Geng, S.; Shi, P.; Song, M.; Zong, N.; Zu, J.; Zhu, W. Diversity of vegetation composition enhances ecosystem stability along elevational gradients in the Taihang Mountains, China. *Ecol. Indic.* **2019**, *104*, 594–603. [CrossRef]
122. De Pascale, S.; Arena, C.; Aronne, G.; De Micco, V.; Pannico, A.; Paradiso, R.; Rouphael, Y. Biology and crop production in Space environments: Challenges and opportunities. *Life Sci. Space Res.* **2021**, *29*, 30–37. [CrossRef] [PubMed]
123. Zhang, X.; Zhang, Z.; Li, Z.; Li, M.; Wu, H.; Jiang, M. Impacts of Spartina alterniflora invasion on soil carbon contents and stability in the Yellow River Delta, China. *Sci. Total Environ.* **2021**, *775*, 145188. [CrossRef]
124. Wawrzyczek, J.; Lindsay, R.; Metzger, M.J.; Quetier, F. The ecosystem approach in ecological impact assessment: Lessons learned from windfarm developments on peatlands in Scotland. *Environ. Impact Assess. Rev.* **2018**, *72*, 157–165. [CrossRef]
125. Fu, C.; Xu, M. Achieving carbon neutrality through ecological carbon sinks: A systems perspective. *Green Carbon* **2023**, *1*, 43–46. [CrossRef]
126. Tan, Z.Q.; Yao, J.; Gong, L.Q.; Li, Y.L.; Zhang, Q.; Wang, X.L.; Wan, R.R.; Li, B. Multidirectional Shifts in Hydrological Connectivity Result From Hydraulic Barriers. *Water Resour. Res.* **2023**, *59*, e2022WR033617. [CrossRef]
127. Liu, Q.; Jiang, Y.; Huang, X.; Liu, Y.; Guan, M.; Tian, Y. Hydrological conditions can change the effects of major nutrients and dissolved organic matter on phytoplankton community dynamics in a eutrophic river. *J. Hydrol.* **2024**, *628*, 130503. [CrossRef]
128. Prescott, C.E.; Vesterdal, L. Decomposition and transformations along the continuum from litter to soil organic matter in forest soils. *For. Ecol. Manag.* **2021**, *498*, 119522. [CrossRef]
129. Li, S.; Wu, M.; Jia, Z.; Luo, W.; Fei, L.; Li, J. Influence of different controlled drainage strategies on the water and salt environment of ditch wetland: A model-based study. *Soil Till Res.* **2021**, *208*, 104894. [CrossRef]
130. Ji, M.; Dong, R.; Zhang, J.; Shi, X.; Wang, Y.; Huang, Q.; Qu, D.; Wang, Y. Diversity and environmental determinants of aquatic plants across China. *Hydrobiologia* **2024**, *851*, 3453–3469. [CrossRef]
131. Zheng, H.; Liu, D.; Yuan, J.; Li, Y.; Li, J.; Miao, Y.; Chen, Z.; He, T.; Ding, W. Wetland restoration after agricultural abandonment enhances soil organic carbon efficiently by stimulating plant- rather than microbial-derived carbon accumulation in Northeast China. *Catena* **2024**, *241*, 108077. [CrossRef]

water

MDPI

Article

Suitability Evaluation of Ecological Restoration Relying on Water Resources in an Agro-Pastoral Transition Zone: A Case Study of Zhangbei, Zhangjiakou, Northern China

Jin-Jie Miao [1,2,3], Yi-Hang Gao [1,2,3,*], Ying Zhang [4], Xue-Sheng Gao [1,2,3], Dan-Hong Xu [1,2,3], Jun-Quan Yang [1,2,3], Wei Wang [1,2,3] and Hong-Wei Liu [1,2,3,*]

1 Tianjin Center, China Geological Survey (North China Center of Geoscience Innovation), Tianjin 300170, China; tjmiaojj@163.com (J.-J.M.); xuesheng.8201@163.com (X.-S.G.); s454650073@126.com (D.-H.X.); dap-yangjunquan@163.com (J.-Q.Y.); wangwei_wangwei@126.com (W.W.)
2 Xiongan Urban Geological Research Center, China Geological Survey, Tianjin 300170, China
3 Tianjin Key Laboratory of Coast Geological Processes and Environmental Safety, Tianjin 300170, China
4 Tianjin Institute of Geological Survey, Tianjin 300191, China; yingzhang@tianjin.cgs.gov.cn
* Correspondence: gaoyihang1988@163.com (Y.-H.G.); liuhenry022@163.com (H.-W.L.)

Abstract: (1) Background: Ecological restoration is crucial to improve ecological functions and optimize its security patterns. The Zhangbei of Zhangjiakou, a typical agro-pastoral transition zone, was studied as an example to conduct ecological restoration suitability evaluation in northern China. (2) Methods: suitability of ecological restoration in Zhangbei was assessed by both single factor analysis and comprehensive factor analysis, which were based on the data of regional water resources, ecosystem service function, and ecosystem sensitivity obtained from a high-precision environmental survey. (3) Results and conclusions: The results show that in Zhangbei County, areas classified as important and extremely important for ecosystem service functions account for 50.32%, ecologically sensitive and highly sensitive areas represent 5.95%, and regions designated as important and extremely important for ecological protection cover 52.70%. Furthermore, ecological restoration of Zhangbei was divided into four ecological restoration zones: agro-forest–wetland ecological restoration and soil erosion control zone, agro-forest–wetland ecological restoration and water conservation zone, forest–grassland soil erosion and soil–water conservation zone, and mountain forest conservation and biodiversity maintenance zone. The study can be a scientific case study for local ecosystem restoration and conservation. In the future, this study will further explore multi-source data fusion, the establishment of a multi-scale evaluation system, and the trade-off analysis between conservation and development.

Keywords: ecological restoration zoning; water resources; dominant ecological function; Zhangbei

1. Introduction

Rapid development of the regional economy encouraged steadily accelerated urban expansion, which caused noticeable ecological issues due to overuse of various spatial resources [1]. The sustainable growth of regional ecological security has been impacted by ongoing human activity [2,3]. Thus, targeted regional ecological protection and restoration is desired to reduce the current severe trend of ecological and environmental deterioration [4,5].

Foreign studies in this field originated from early 20th century land evaluation theories [6], followed by the enactment of a series of laws, regulations, and "restoration programs" [7–12]. In the United States, the "Land Suitability Classification" method, initially developed for agricultural productivity assessment, was later extended to ecological restoration, emphasizing climate adaptability and species selection [13]. European scholars prioritize ecosystem service value assessment, using outcome benefit analysis to determine restoration priorities. For instance, Germany integrates carbon sequestration functions into evaluation criteria [14]. Japan developed comprehensive models combining geographic positioning and ecological functions, promoting the optimization of agroforestry systems [15]. In recent years, global-scale ecological restoration research shifted toward interdisciplinary approaches, such as quantifying ecological carrying capacity using the the InVEST model and enhancing the feasibility of restoration plans through community engagement mechanisms [16].

Since the 1950s, more and more ecological restoration projects were carried out in China, and abundant theoretical methodologies [17–20], technical specifications [21], and real cases [22–27] were accumulated gradually. Cai et al. [28] introduced a way of ecological restoration regionalization based on dominant functional features, such as regional land, ecological, and economic functions. These methods were based on the fundamental theories of land planning and ecological civilization construction, such as "life community theory" and "suitability and management." Yu et al. [29] proposed regional land comprehensive remediation measures, which supports the idea of coupling supply and demand, including a whole-process "1 + 1 + X" global remediation land space planning system, multi-disciplinary natural resource value potential, and global remediation performance evaluation methods. Overall, current evaluation methods on ecological restoration mostly rely on individual parameters, such as ecological security patterns, ecological sensitivity, and ecosystem service function [30,31]. Nevertheless, there were few studies based on two or more parameters, and they mostly carried on within a large scale. It is necessary to attempt the comprehensive management at local regional levels with multiple parameters.

The agro-pastoral ecotone refers to a transitional zone between traditional agriculture and animal husbandry, widely distributed in northern China and serving as a critical ecological barrier for the region [28]. However, due to its inherently fragile natural environment and the rapid growth of the population and economy, the natural resources and ecological conditions in this area have been increasingly deteriorating [32–38]. For instance, the Grain for Green program (conversion of cropland to forest and grassland) implemented on the Loess Plateau failed to adequately account for regional differences in precipitation and soil conditions, resulting in suboptimal ecological restoration outcomes in certain areas. Zhangjiakou's Zhangbei County, a typical agro-pastoral ecotone in northern China, simultaneously fulfills multiple environmental service functions, including biodiversity conservation, carbon sequestration, agricultural and pastoral resource production, and ecotourism. Such areas often lie in zones where ecological fragility overlaps with high service value, and their restoration outcomes will directly shape the regional ecological security pattern. Since the year 2000, the implementation of the Bashang Project for Returning Farmland to Forests and Grasslands positively impacted local ecological restoration [35]. However, conflicts persist between ecological preservation and economic development, urgently necessitating research on ecological protection and restoration. This research should optimize regional ecological security spatial patterns, comprehensively address environmental issues, and enhance ecosystem service functions in the area.

Therefore, this study selects Zhangbei County as the research area to establish a comprehensive evaluation system integrating water conservation, soil and water retention, biodiversity maintenance, and windbreak sand fixation functions. A progressive "function-sensitivity-protection level" framework is developed, utilizing high-precision environmental geological data and provincial "dual evaluation" (ecological–environmental and agricultural–production suitability assessments) outcomes for multi-scale collaboration. Sensitivity classification criteria are formulated based on hydraulic and wind erosion intensity. Through a "preliminary judgment-coordination-correction" verification mechanism, multi-source data, including ecological corridors, geomorphic integrity, and protected area planning, are integrated to create a refined ecological protection zoning methodology.

This approach provides an integrated "assessment-protection-restoration" technical solution for ecologically fragile agro-pastoral ecotones in northern China, addressing the fragmentation of traditional planning elements. It precisely supports the optimization of Zhangbei's ecological redline delineation and territorial spatial governance, offering a replicable paradigm for constructing ecological security patterns in similar counties across China. The methodology innovatively bridges ecological sensitivity analysis with practical conservation needs, enabling data-driven decision-making for balancing ecological resilience and sustainable development in vulnerable regions.

2. Materials and Methods

2.1. The Study Area

Zhangbei is situated in the Bashang region of the Inner Mongolia Plateau in northwestern Hebei Province (Figure 1). The terrain is predominantly characterized by plains and hills, with elevations sloping downward toward the central area and rising in the southern and northern parts [39]. The altitude is 1200~2000 m above sea level, and the climate belongs to the temperate continental grassland zone, with approximately 400 mm of annual precipitation. The Anguli Lake Basin hosts an extensive network of 25 rivers, with a total length of 793 km. Most rivers originate from the hilly uplands of the dam and flow from south to north. These waterways are typically short, with gentle gradients, and most terminate into onshore basins, and a few flow into inland depressions. The ground surface is highly unstable here due to its thick loose sedimentary layers of quaternary, particularly widespread aeolian sand and loess, which causes the land to become exceptionally vulnerable from wind and water erosion. As a result, desertification and degradation of forest and grassland ecosystems are chiefly environmental challenges in this region.

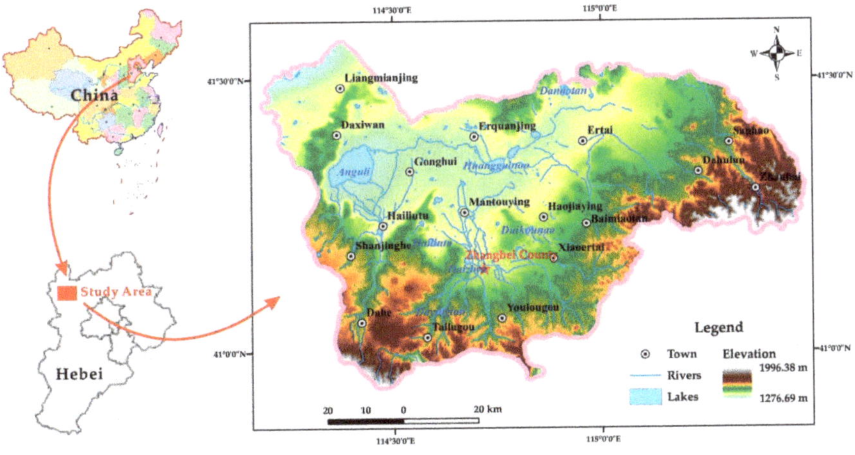

Figure 1. The location of study area.

2.2. Data Sources

2.2.1. Basic Data

The selection of data sources in this study adheres to principles of scientific rigor, authoritative credibility, and spatiotemporal consistency. The specific data elements and their selection rationale are detailed below (Table 1):

Table 1. Main data applied in this article.

Data	Description	Resolution Ratio/km	Data Sources
Elevation data set	DEM taster data	0.03	Geo-spatial data cloud
NDVI data set	Raster	1	MOSIS
NPP data set	Raster	1	GLASS dataset
Soil data set	Raster	1	Geo-graphic remote sensing ecological network platform
Meteorological and climate data sets	Raster/text	1	China meteorological science data

Digital elevation model (DEM) data were employed to derive slope and aspect parameters (sourced from the Geographic Spatial Data Cloud Platform, Chinese Academy of Sciences [40], whose 30 m resolution ensures precision for county scale analysis). Normalized difference vegetation index (NDVI) values were obtained from MODIS satellite products (MOD13Q1), with NASA's global validation network calibration enabling accurate reflection of interannual vegetation coverage variations [41]. Net primary productivity (NPP) data were selected from the GLASS dataset, which mitigates single-sensor errors through multi-source remote sensing integration [42]. Soil characteristic parameters were derived from the China subset (V1.1) of the Harmonized World Soil Database (HWSD), containing critical 0–100 cm soil layer organic matter metrics after localization calibration by China Agricultural University researchers, meeting ecological carrying capacity modeling requirements. Meteorological data were acquired from the China Meteorological Science Data Sharing Service Network, utilizing monthly raster datasets (2000–2020) with 1 km$\times$1 km spatial resolution, whose observation point density (2419 nationwide stations) demonstrates high spatial congruence with the agro-pastoral ecotone study area.

2.2.2. High-Resolution Environmental Geological Survey of Agro-Pastoral Transition Zone

A high-resolution environmental geology survey of the agro-pastoral transition zone in the study area was conducted by field validation, supported by high-resolution remote sensing imagery. Accordingly, land degradation in Zhangbei is mainly due to saline–alkaline desertification, with a total area of 314.60 km^2 in 2020, including 216.28 km^2 with light degree, 39.55 km^2 with moderate degree, and 58.77 km^2 with severe degree. The total grassland of this area is 753.07 km^2, including 62.40 km^2 as heavily salinized, 211.73 km^2 as salinized grassland, and 582.33 km^2 usable grassland.

2.3. Evaluation of Ecosystem Service Function Importance

To safeguard regional ecological security, this study employs land use suitability and resource-carrying capacity frameworks. Namely, it evaluates the ecological functions (e.g., biodiversity maintenance, windbreak and sand fixation, and water–soil conservation) and the ecological sensitivity (e.g., soil erosion and land desertification risks) of the study area. The importance of the degree of ecosystem services functions is generated from spatially integrating the evaluation results of four critical items, including water conservation, soil retention, biodiversity preservation, and windbreak sand fixation.

2.3.1. Evaluation of Water Conservation Function Importance

Water conservation refers to the capacity of ecosystems (e.g., forests, grasslands) to regulate hydrological cycles by intercepting, infiltrating, and storing precipitation through their structural features. Following the Ecological Conservation Redline Delineation Guidelines [43], this study evaluates water conservation capacity by a water budget decomposition model that partitions precipitation and evapotranspiration (Figure 2). Generally, grid-based water conservation values are ranked from highest to lowest. The next step includes calculating the cumulative values and identifying thresholds at 50% and 80% of the cumulative value, and finally, classifying water conservation importance into three degrees: critically important, important, and generally important, based on the thresholds identified.

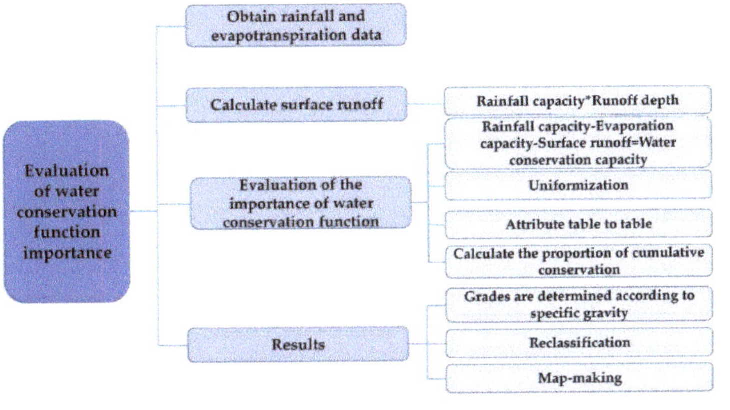

Figure 2. Evaluation processes of water conservation function importance.

2.3.2. Evaluation of Soil and Water Conservation Function Importance

Soil and water conservation refers to the capacity of ecosystems (e.g., forests, grasslands) to mitigate soil erosion from water flow by their structural and functional processes, serving as a critical regulatory ecosystem service [43]. Slopes are generally gentle in Zhangbei, so slope gradients were classified into three degrees from the steepest to the gentlest by the natural breaks method (Table 2). For example, the critically important zones are characterized with the steepest slopes and vegetation coverage greater than or equal to 60%; and important zones are characterized with generally steep slopes and vegetation coverage greater than or equal to 20%.

Table 2. Thresholds for soil and water conservation functions importance.

Slope Gradient	Forest, Scrub and Grassland Coverage More Than 60%	Forest, Scrub and Grassland Coverage Is 20–60%	Forest, Scrub and Grassland Cover Less Than 20%
Steep	5	3	1
Relatively steep	3	3	1
Gradual	1	1	1

2.3.3. Evaluation of Biodiversity Maintenance Function Importance

The biodiversity maintenance function refers to an ecosystem's capacity to sustain genetic, species, and ecosystem diversity, and it is one of its most critical functions. This study employs the net primary productivity (NPP) method to assess biodiversity maintenance importance by the biodiversity maintenance service capacity index (Sbio). At first, grid-based service values are ranked from highest to lowest and then the cumulative values are computed. Next, critical thresholds at 50% and 80% of the total cumulative ecosystem service values are identified. Finally, biodiversity maintenance is classified into three categories: critically important, important, and generally important.

$$S_{bio} = NPP_{mean} \times F_{pre} \times F_{tem} \times (1 - F_{alt}) \quad (1)$$

where S_{bio} is the biodiversity maintenance service capability index; NPP_{mean} is the multi-year average of vegetation; F_{pre} is the normalized precipitation factor; F_{tem} is the normalized temperature factor; and F_{alt} is the normalized elevation factor.

2.3.4. Evaluation of Windbreak and Sand Fixation Function Importance

Windbreak and sand fixation function refers to the capacity of ecosystems (e.g., forests, grasslands) to mitigate soil erosion from wind by their structural and functional processes, serving as a vital regulatory ecosystem service. This study assesses the importance of the windbreak sand fixation function from the windbreak and sand fixation service capacity index (SWS). Accordingly, it can be classified into critically important zones and important zones. The critically important zones are the areas with annual precipitation < 400 mm, strong wind speeds, gentle slopes (<5°), and sandy soils, also including built-up land and gobi desert with vegetation cover <20%. The important zones are the areas with precipitation $\leq$ 60% of regional averages, moderate wind speeds, steeper slopes (>5°), and saline soils, also including farmland and water bodies vegetation cover $\geq$10%. The formula for the windbreak and sand fixation service capacity index (SWS) is calculated as follows:

$$S_{WS} = NPP_{mean} \times K \times F_q \times D \quad (2)$$

$$F_q = \frac{1}{100}\sum_{i=1}^{12} u^3 \left\{ \frac{ETP_i - P_i}{ETP_i} \right\} \times d \quad (3)$$

$$ETP_i = 0.19(20 + T_i) \times (1 - r_i) \quad (4)$$

$$u_2 = u_1 (z_2/z_1)^{1/7} \quad (5)$$

$$D = 1/cos\theta \quad (6)$$

where S_{WS} is the windbreak and sand fixation service capacity index, K is the soil erodibility factor, F_q is the multi-year average climate erosivity factor, D is the surface roughness factor, u is the monthly average wind speed at a height of 2 m, and u_1, u_2 is the wind speed at a height of z_1, z_2, ETP_i is the monthly potential evaporation (mm), P_i is the monthly precipitation (mm), d is the number of days in the month, T_i is the monthly average temperature (°C), r_i is the monthly average relative humidity (%), and θ is the slope (°).

2.4. Evaluation of Ecological Sensitivity

Soil erosion sensitivity and land desertification sensitivity were classified into two levels: highly sensitive and sensitive, with the highest sensitivity level of all factors determining the final ecological sensitivity rating. According to high-resolution environmental geological survey data from the agro-pastoral transition zone in Zhangbei, the area with severe and extremely severe hydraulic/wind erosion intensities was classified into highly sensitive; the area with intense and moderate erosion levels was classified into sensitive.

2.5. Evaluation of Ecological Protection Importance

The preliminary delineation of the critically important zones of ecological protection was determined by overlaying and integrating areas both with critically important in ecosystem service function and highly sensitive, identified by ecological sensitivity evaluations. Afterwards, the results were aligned with the province evaluation result to ensure consistency in regional ecological planning frameworks. Further refinements were applied with consideration of protected area plans: designated conservation zones in Zhangbei; ecological corridors: habitats of rare and endangered species; and geoenvironmental integrity: landscape features, geographical coherence, and ecosystem continuity. Key references for refining the preliminary critically important zones included the following: ecological core areas (3 sites); scenic landscapes (1 site); geoparks (1 site); water source protection zones (1 site); and inland wetlands (Figure 3).

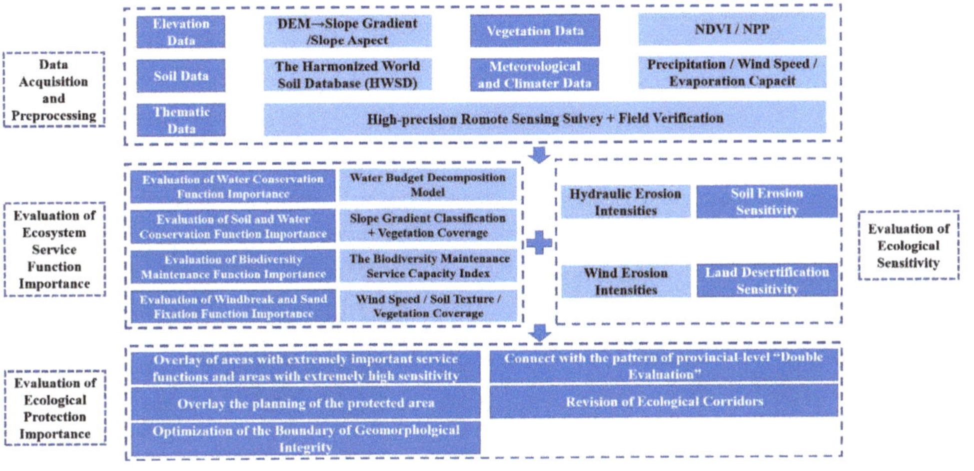

Figure 3. Flowchart of the evaluation of the importance of ecological protection.

2.6. Principles for Naming Ecological Restoration Zones

The ecological restoration zoning framework for Zhangbei was developed with single-factor analysis and comprehensive multi-factor analysis by GIS spatial clustering techniques to identify regional advantages and classify land ecosystem types. There were two conventions for the classification of the ecological restoration zone: ecosystem subsystem and dominant ecological function/sensitivity/environmental issue. The ecosystem subsystem included forest, grassland, farmland, and urban areas; the environmental features included drinking water source protection, wetland degradation, and critical geological heritage sites; and the ecological functions/sensitivities included water conservation, soil retention, biodiversity preservation, wind erosion control, soil erosion sensitivity, and land desertification. The final nomenclature prioritizes regionally dominant or representative characteristics, ensuring alignment with local ecological priorities and restoration objectives.

3. Results and Discussion

Although the core methodology of this study relies on remote sensing technology, stratified sampling surveys were conducted in key sensitive areas of the research region, collecting soil samples, groundwater, and surface water samples. These data were cross-verified with ground-based observations from references [36,44–46], leading to the following results.

3.1. Evaluation Results of Ecological Service Function Importance

Accordingly, the water conservation function, soil and water conservation function, biodiversity maintenance function, and windbreak and sand fixation function of Zhangbei were assessed (Figure 2, Table 3), and then a composite ecological service importance classification integrating from four functional evaluations was generated (Figure 4).

Table 3. Evaluation results of ecological functions in Zhangbei.

Function	Extremely Important		Important		Generally Important	
	Area (km^2)	Proportion (%)	Area (km^2)	Proportion (%)	Area (km^2)	Proportion (%)
Water conservation	445.53	10.31	1951.45	45.17	1923.71	44.52
Soil and water conservation	302.86	7.01	2957.12	68.44	1060.71	24.55
Biodiversity maintenance	495.72	11.47	1004.93	23.26	2820.04	65.27
Windbreak and sand fixation	1753.13	40.58	2067.67	47.85	499.89	11.57

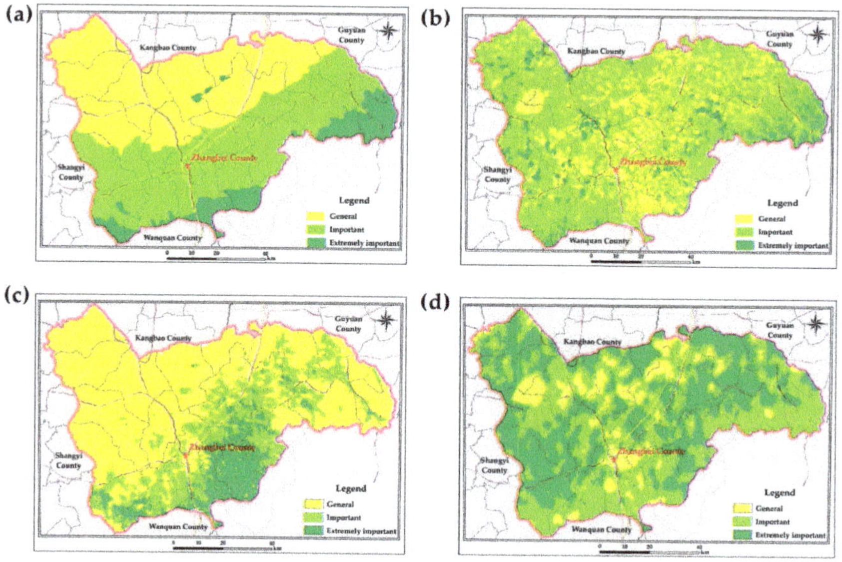

Figure 4. Ecological service function importance in Zhangbei. (**a**) Evaluation of water conservation function importance; (**b**) evaluation of soil and water conservation function importance; (**c**) evaluation of biodiversity maintenance function importance; and (**d**) evaluation of windbreak and sand fixation function importance.

3.1.1. Evaluation Results of Water Conservation Function Importance

Water conservation refers to the capacity of ecosystems (e.g., forests, grasslands) to regulate hydrological cycles through their structural and functional interactions with water. This involves intercepting, infiltrating, and storing precipitation while modulating water flow and distribution via evapotranspiration. It includes mitigating surface runoff, replenishing groundwater, moderating seasonal river discharge fluctuations, regulating floodwaters and sustaining baseflow during dry periods, and maintaining water quality. In Zhangbei, according to the evaluation, the critically important zones for water conservation cover approximately 445.53 km^2 (10.31% of the total area; Table 3), and are primarily concentrated in southern Zhangbei, and surround the Huanggainao Lake in the north (Figure 4a). Important zones cover 1951.45 km^2 (45.17%), while generally important zones cover 1923.71 km^2 (44.52%).

3.1.2. Evaluation Results of Soil and Water Conservation Function Importance

Soil and water conservation refers to the capacity of ecosystems (e.g., forests, grasslands) to mitigate soil erosion caused by water flow. In the study area, due to an average elevation of 1400~1600 m and favorable hydrothermal conditions, vegetation is dominated by forestland and pasture (Figure 4b), resulting in robust soil retention capabilities. Consequently, both the critically important and the important zones cover 75.45% of the total area (Table 3), and the critically important zones span 302.86 km^2 (7.01%) and the important zones span 2957.12 km^2 (68.44%). The generally important zones span 1060.71 km^2 (24.55%).

3.1.3. Evaluation Results of Biodiversity Maintenance Function Importance

Biodiversity maintenance function encompasses the role of an ecosystem in preserving genetic, species, and ecosystem diversity. In the study area, the critically important zones for biodiversity conservation cover 495.72 km^2 (11.47%; Table 3), concentrated around the Xiao'ertai town in eastern Zhangbei (Figure 4c). The important zones cover 1004.93 km^2 (23.26%), while generally important zones dominate the northwestern region, accounting for 2820.04 km^2 (65.27%).

3.1.4. Evaluation Results of Windbreak and Sand Fixation Function Importance

Windbreak and sand fixation function refers to ecosystems' capacity to reduce soil erosion driven by wind. In Zhangbei, the critically important zones for windbreak sand fixation span 1753.13 km^2 (40.58% of the total area), distributed ubiquitously across the region (Figure 4d, Table 3). The important zones span 2067.67 km^2 (47.85%), and the generally important zones only span 499.89 km^2 (11.57%).

The result, from spatially integrating these four key functions, reveals a distinct north–south gradient in ecosystem service importance (Figure 5, Table 4). The critically important zones cover 203.69 km^2 (4.71%). The important zones span 1907.77 km^2 (45.61%) and predominantly cluster in southern Zhangbei, while the generally important zones dominate northern Zhangbei and span 2146.23 km^2 (49.67%).

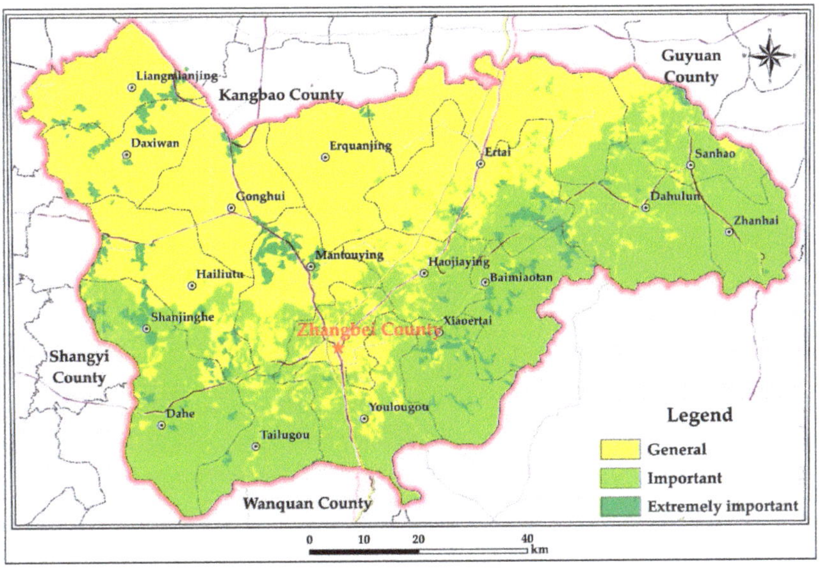

Figure 5. Evaluation of ecosystem service function importance in Zhangbei.

Table 4. Evaluation of ecosystem service function importance in Zhangbei.

Function	Extremely Important		Important		Generally Important	
	Area (km^2)	Proportion (%)	Area (km^2)	Proportion (%)	Area (km^2)	Proportion (%)
Ecosystem service function	203.69	4.71	1970.77	45.61	2146.23	49.67

3.2. *Evaluation Results of Ecological Sensitivity*

Overall, the moderate ecological sensitivity is predominant in the study area (Figure 6, Table 5), covering 4063.34 km^2 (over 94% of the total area). The highly sensitive zones account for 179.44 km^2 (4.15%), concentrated in wetlands such as Anguli Nao, Hailiutu Lake, and Dayingtan Lake. The sensitive zones span 77.91 km^2 (1.80%).

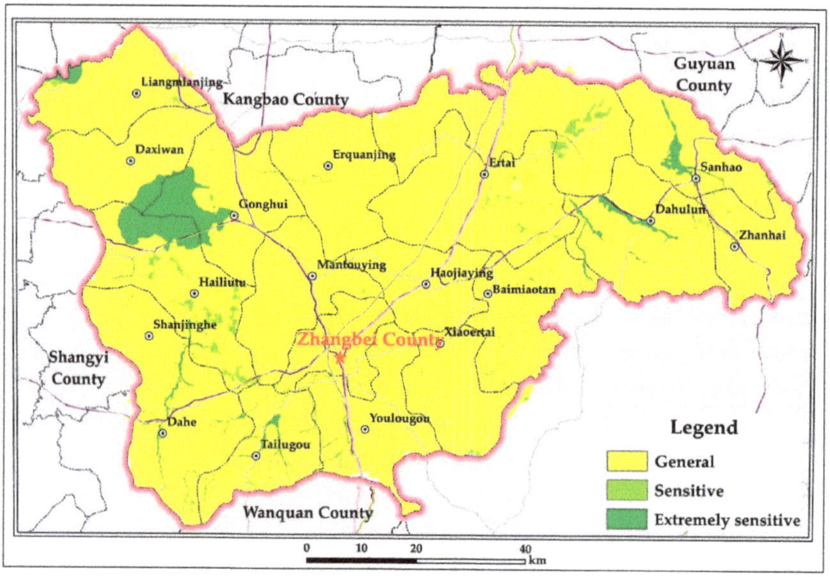

Figure 6. Evaluation of ecological sensitivity of Zhangbei.

Table 5. Summary table of ecological sensitivity evaluation results of Zhangbei.

Function	Highly Sensitive		Sensitive		Moderate	
	Area (km^2)	Proportion (%)	Area (km^2)	Proportion (%)	Area (km^2)	Proportion (%)
Ecological sensitivity	179.44	4.15	77.91	1.80	4063.34	94.04

3.3. *Evaluation Results of Ecological Protection Importance*

For ecological conservation, the majority of Zhangbei is classified as the important zones (1943.86 km^2; 44.99%) and the generally important zones (2043.79 km^2; 47.30%). The extremely important zones of ecological protection cover only 333.04 km^2 (7.71%), primarily clustered around wetlands, such as Angulinao, Hailiutu Lake, and Dayingtan Lake (Figure 7, Table 6).

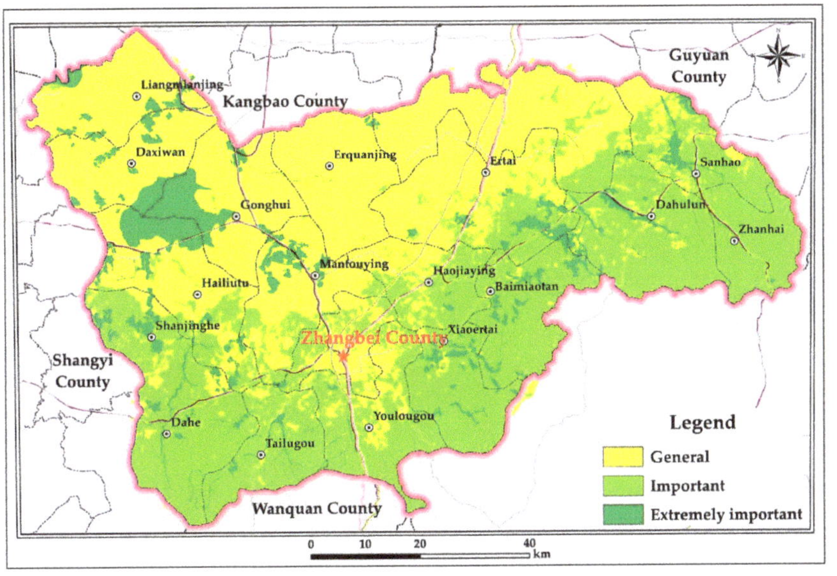

Figure 7. Evaluation of ecological protection importance in Zhangbei.

Table 6. Evaluation of ecological protection importance in Zhangbei.

Function	Extremely Important		Important		Generally Important	
	Area (km^2)	Proportion (%)	Area (km^2)	Proportion (%)	Area (km^2)	Proportion (%)
Ecological protection	333.04	7.71	1943.86	44.99	2043.79	47.30

3.4. Results of Ecological Restoration Zoning

According to the evaluation results of the ecosystem service importance, ecological sensitivity importance, and ecological protection importance in Zhangbei and adhering to the aforementioned zoning and naming principles, the study area is divided into four ecological restoration zones (Figure 8) as follows:

- Agro-forest–wetland ecological restoration and soil erosion control zone

This zone is dominated by forest and grassland, and is mainly located in northwestern Zhangbei, such as at Liangmianjing, Daxiwan, Gonghui, northwestern Hailiutu, and northern Mantouying. Anguli Nao Lake, in this zone, is one of the largest plateau lakes in northern China, but now it is severely deteriorating due to abrupt environmental change and overexploitation, which also exacerbates soil erosion. Thus, strategies including forest conservation and farmland-to-forest conversion are applied to rehabilitate degraded ecosystems, which helps the damaged ecosystem gradually recover.

- Agro-forest–wetland ecological restoration and water conservation zone

This zone is dominated by farmland, forests, and wetlands, and mainly located in southwest Zhangbei, such as at Shanjinghe, Dahe, Tailugou, southern Hailiutu, southern Mantouying, and western Youlougou. Thus, strategies in this zone including rational agricultural resource allocation, site-specific afforestation to restore forest cover, and wetland rehabilitation in degraded areas are applied to enhance hydrological regulation and biodiversity.

- Forest–grassland soil erosion and soil–water conservation zone

This zone is dominated by dense forests, and is mainly located in northern Zhangbei, such as at Erquanjing, Shagou, and northwestern Er'tai town. In this zone, intensive land use led to sparse vegetation, weak root systems, and severe soil erosion exacerbated by rainfall and human activities. Thus, strategies in this zone include establishing ecological shelterbelt systems, and strengthening vegetation protection and soil conservation are applied to curb erosion and boost ecosystem resilience.

- Mountain forest conservation and biodiversity maintenance zone

This zone is dominated by dense forests, and is mainly located in southeastern Zhangbei, such as at Xiao'ertai, Baimiaotan, southeastern Er'tai town, southern Yuzhouying, Dahulun, Sanhao, and Zhanhai town. In this zone, it is covered by dense forests, diverse vegetation, and rugged terrain, which is critical for soil–water conservation and climate regulation. Thus, strategies in this zone, including banning deforestation and unregulated land clearing, and implementing graded forest protection, differentiated management, and compensation mechanisms are applied to stabilize soil retention and sustain forest ecosystem health.

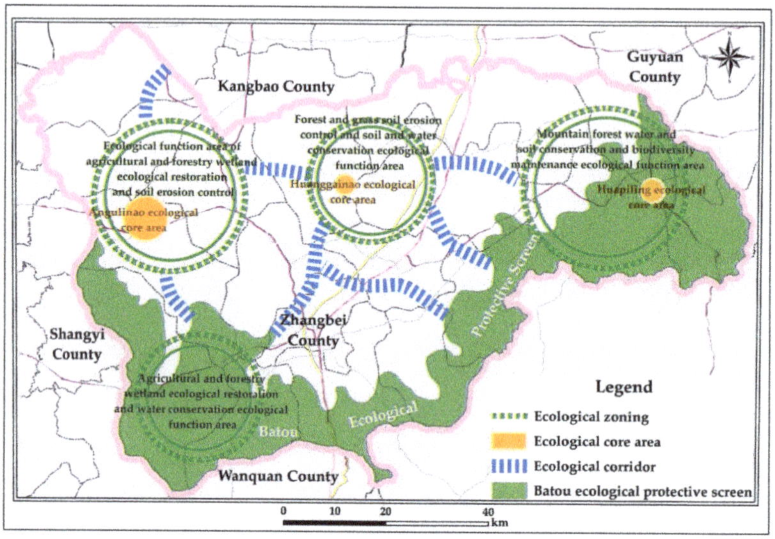

Figure 8. Ecological restoration zoning map of Zhangbei.

4. Conclusions

This paper studied Zhangbei, Zhangjiakou in order to assess its importance of ecosystem function, ecological sensitivity function, and ecological conservation function, according to its current ecological conditions. Based on these evaluations, it was divided into targeted ecological restoration zones, and concludes the following:

- The critically important zones of water conservation function are mainly distributed in southern Zhangbei and surround Huanggainao Lake in the north. The critically important zones of soil and water conservation function are mainly distributed in regions with abundant forest and grassland coverage. The critically important zones of water conservation function are mainly distributed in southern Zhangbei and surround the Huanggainao Lake in the north. The critically important zones of soil and water conservation are distributed in regions with abundant forest and grassland coverage. The critically important zones of biodiversity maintenance functions are mainly distributed around Xiaoertai Lake and its vicinity. The critically important zones of windbreak and sand fixation function are ubiquitously distributed throughout the country. The highly sensitive zones of ecological sensitivity, covering 4.15% of this area, are mainly distributed around wetland such as Anguli Nao. The important and generally important zones of ecological protection exhibit a distinct north–south spatial gradient. The critically important conservation zones with only 333.04 km^2 (7.71%) predominantly overlap with wetland systems.

- According to the evaluation of ecosystem service importance, ecological sensitivity importance, and ecological protection importance in Zhangbei, the study area is divided into four ecological restoration zones: agro-forest–wetland ecological restoration and soil erosion control zone, agro-forest–wetland ecological restoration and water conservation zone, forest–grassland soil erosion and soil–water conservation zone, and mountain forest conservation and biodiversity maintenance zone. Each zone was assigned tailored restoration measures, such as afforestation, wetland rehabilitation, and erosion control infrastructure, to address region-specific ecological challenges.
- Compared to conventional approaches, this study prioritizes dominant ecological functions as the cornerstone of ecological restoration zoning in Zhangbei. By anchoring both the conceptual framework and practical implementation in these functions, the methodology clarifies pathways for integrated territorial space management and establishes a scientifically grounded basis for local ecological restoration and spatial governance.
- Next, research will develop advanced multi-source data fusion to enhance DEM/NDVI spatiotemporal resolution, improve real-time vegetation/soil monitoring, conduct multi-scale evaluations and ecological simulations analyzing spatial heterogeneity impacts, and establish ecology–economy models to evaluate synergies between conservation and agricultural/energy development.

Author Contributions: Conceptualization, J.-J.M. and Y.-H.G.; methodology, Y.-H.G.; software, Y.-H.G. and D.-H.X.; validation, J.-J.M.; investigation, Y.-H.G., J.-J.M., X.-S.G. and W.W.; resources, J.-Q.Y.; data curation, Y.-H.G. and D.-H.X.; writing—original draft preparation, Y.-H.G.; writing—review and editing, J.-J.M., Y.Z. and H.-W.L.; visualization, Y.-H.G.; supervision, J.-J.M. and J.-Q.Y.; project administration, J.-Q.Y. and W.W.; funding acquisition, J.-J.M. All authors have read and agreed to the published version of the manuscript.

Funding: This research was funded by "the geological safety physical examination and risk assessment in Tianjin van (Grant no. DD20243452); the Monitoring and evaluation of resource and environment Carrying capacity of the Beijing-Tianjin-Hebei Collaborative Development Area and Xiongan New Area (Grant no. DD20221727).

Data Availability Statement: This study does not involve any patented technologies that have been applied for or are under review. All experimental data and methodologies were conducted in accordance with publicly available scientific research protocols.

Conflicts of Interest: The authors declare that they have no conflicts of interest.

References

1. Fu, B.J. Trends and priority areas in ecosystem research of China. *Geogr. Res.* **2010**, *29*, 383–396.
2. Lovell, S.T.; Taylor, J.R. Supplying urban ecosystem services through multifunctional green infrastructure in the United States. *Landsc. Ecol.* **2013**, *28*, 1447–1463. [CrossRef]
3. Yang, S.; An, Y.; Wang, P.; Ma, L.; Hu, F.; Sun, Q. Study of ecological red-line zones in Guizhou Chishui River basin. *Resour. Environ. Yangtze Basin* **2015**, *24*, 1405–1411.
4. Hou, P.; Yang, M.; Zhai, J.; Liu, X.; Wan, H.; Li, J.; Cai, M.; Liu, H. Discussion about natural reserve and construction of national ecological security pattern. *Geogr. Res.* **2017**, *36*, 420–428.
5. Li, S.; Zhai, H. The comparison study on forestry ecological projects in the world. *Acta Ecol. Sin.* **2002**, *22*, 1976–1982.
6. Ehrenfeld, J.G. Defining the limits of restoration: The need for realistic goals. *Restor. Ecol.* **2000**, *8*, 2–9. [CrossRef]

7. Herbertson, A.J. The major natural regions: An essay in systematic geography. *Geogr. J.* **1905**, *25*, 300. [CrossRef]
8. Palmer, M.A.; Bernhardt, E.S.; Allan, J.D.; Lake, P.S.; Alexander, G.; Brooks, S.; Carr, J.; Clayton, S.; Dahm, C.N.; Follstad Shah, J.; et al. Standards for ecologically successful river restoration. *J. Appl. Ecol.* **2005**, *42*, 208–217. [CrossRef]
9. Mace, G.M. Whose conservation? *Science* **2014**, *345*, 1558–1560. [CrossRef]
10. Geldmann, J.; Joppa, L.N.; Burgess, N.D. Mapping change in human pressure globally on land and within protected areas. *Conserv. Biol.* **2014**, *28*, 1604–1616. [CrossRef]
11. Higgs, E.; Harris, J.; Murphy, S.; Bowers, K.; Hobbs, R.; Jenkins, W.; Kidwell, J.; Lopoukhine, N.; Sollereder, B.; Suding, K.; et al. On principles and standards in ecological restoration. *Restor. Ecol.* **2018**, *26*, 399–403. [CrossRef]
12. Madrona Conservation Initiative. Prioritizing Aquatic Conservation and Restoration in the Pacific Northwest (Techniacal Report/EB/OL). 2022. Available online: https://www.madrona.org/reports/aquatic-conservation-pacific-2022 (accessed on 3 May 2025).
13. Lathrop, R.B. Applying GIS and landscape ecological principles to evaluate land conseration alereratives. *Landsc. Urban Plan.* **1988**, *41*, 27–41. [CrossRef]
14. Stoorvogel, J. Intergration of computer-based models and GIS tools to evaluate altermative land-use scenatries as part of an agricultural systems analysis. *Agric. Syst.* **1995**, *49*, 353–367. [CrossRef]
15. Ayanu, Y.Z.; Conrad, C.; Nauss, T.; Wegmann, M.; Koellner, T. Quantifying and mapping ecosystem services supplies and demands: A review of remote sensing applications. *Environ. Sci. Technol.* **2012**, *46*, 8529–8541. [CrossRef] [PubMed]
16. Wang, X.; Zhang, X.; Mu, X.; Zhu, Z. An analysis on the compilation method of ecological restoration planning for territorial space. *Environ. Prot.* **2019**, *47*, 36–38.
17. Fang, C.; Wang, Z.; Liu, H. Exploration on the theoretical basis and evaluation plan of beautiful China construction. *Acta Geogr. Sin.* **2019**, *74*, 619–632.
18. Wang, J.; Ying, L.; Zhong, L. Thinking for the transformation of land consolidation and ecological restoration in the new era. *J. Nat. Resour.* **2020**, *35*, 26–36.
19. Yi, X.; Bai, C.; Liang, L.; Zhao, Z.; Song, W.; Zhang, Y. The evolution and frontier development of land ecological restoration research. *J. Nat. Resour.* **2020**, *35*, 37–52.
20. Lv, J.X.; Li, Q. Research on the policy system and implement path of territorial ecological restoration. *J. Hebei Acad. Sci.* **2023**, *40*, 62–69.
21. Liu, Y.; Shi, P.; Liu, M.; Xu, K.; Zhang, N.; Jiang, P.; Wang, W.; Jiang, Y. Spatial pattern of water conservation function and ecological management suggestion in the catchment area of the upper reaches of Qinhe River in the Yellow River Basin from 1990 to 2020. *Geol. China* **2024**, *51*, 1917–1929.
22. Wang, C.; Hou, P.; Liu, X.; Yuan, J.; Zhou, Q.; Lv, N. Spatiotemporal changes in vegetation cover of the national key ecosystem protection and restoration project areas, China. *Acta Ecol. Sin.* **2023**, *43*, 8903–8916.
23. Hua, Y.; Pan, J.; Du, J.; Li, Y.; Yang, Q.; Xu, S.; Huang, Y. Effects of the long-term ecological restoration in the eutrophic plateau shallow lake—A case study of Dabokou, Lake Dianchi. *J. Lake Sci.* **2023**, *358*, 1549–1561.
24. Zhao, J.; Sa, N.; Fu, X.; Zheng, S.; Wu, G.; He, X.; Lu, Z.; Sang, W. Multi-scale coupling framework and method for mountains-rivers-forests-farmlands-lakes-grasslands ecological restoration projects: A case study of Shule River Basin. *Acta Ecol. Sin.* **2023**, *43*, 3841–3854.
25. Liang, S.; Zhang, J.; Wang, K.; Liu, S. Methodology and application of carbon sink potential assessment for regional ecological conservation and restoration: Based on the research of the first batch of pilots for ecological protection and restoration project of mountains-rivers-forests-farmlands-lakes-grasslands system. *Acta Ecol. Sin.* **2023**, *43*, 3517–3531.
26. Liu, Y.; Peng, J.; Li, G.; Han, W.; Liu, L.; Guan, J.; Ju, X.; Zheng, J. Eco-environmental assessment of Kurustai grassland based on Google Earth Engine. *Chin. J. Ecol.* **2023**, *42*, 2776–2785.
27. Peng, Y.; Zhuang, X.; Zhao, L.; Wang, Z.; Gao, J.; Wang, B.; He, Z. Influence of species choice and tidal flat elevation on the carbon sequestration of early mangrove restoration. *Acta Sci. Nat. Univ. Sunyatseni* **2023**, *62*, 37–46.
28. Cai, H.; Chen, Y.; Zha, D.; Zeng, H.; Shao, H.; Hong, T. Principle and method for ecological restoration zoning of territorial space based on the dominant function. *Trans. Chin. Soc. Agric. Eng. Trans. CSAE* **2020**, *36*, 261–270.
29. Yu, J.; Dong, Y.; Tian, Y.; Hu, Z. Provincial comprehensive land use reorganization based on resource integration, Zhejiang. *Planners* **2021**, *37*, 17–23.
30. Bai, S.; Wu, M.; Yu, Y. Identification of key ecological restoration areas and optimal allocation of vegetation for water conservation in Hanjiang River Basin. *Bull. Soil Water Conserv.* **2023**, *43*, 123–128.
31. Kong, F.; Duan, S.; Xu, C. The construction of ecological security pattern based on ecosystem services and ecological sensitivity: A case study of the Qiantang River Basin. *Acta Ecol. Sin.* **2024**, *44*, 11359–11374.
32. Dong, Q.; Liang, S.; Liu, B. Current situation and developing counter measures of ecological construction of agriculture and husbandry in Hebei Bashang. *J. Anhui Agric. Sci.* **2006**, *34*, 310–312.

33. Fan, Y.; Hu, N.; Ding, S.; Liang, G.F.; Lu, X.L. Progress in terrestrial ecosystem services and biodiversity. *Acta Ecol. Sin.* **2016**, *36*, 4583–4593.
34. Li, L.; Wu, D.; Liu, Y.; Teng, L.; Wu, Z.; Feng, Z. "Double evaluations" of karst area from the perspective of ecological civilization: A case study of Ningyuan in ecologically sensitive area. *J. Nat. Resour.* **2020**, *35*, 2385–2400. [CrossRef]
35. Zhang, X.; Liang, H.; Hu, Y.; Wang, Z. Comprehensive evaluation and prevention and control zoning of ecological risk based on "ecological sensitivity and interference". *Geogr. Geo-Inf. Sci.* **2020**, *36*, 112–118.
36. Su, H.; Wang, F.; Chen, H.Y.; Liu, X.; Wei, Y.; Luo, Y.; Chen, L.; Bai, M. Planning and practice of ecological protection and restoration project of "mountain-river-forest-farmland-lake-grassland" in Lishui City, Zhejiang Province. *J. Environ. Eng. Technol.* **2022**, *12*, 224–231.
37. Jiang, W.; Meng, L.; Liu, F.; Liu, H.; Zhang, J.; Ning, H. Research on exploitation, utilization and environmental quality of groundwater in Zhangjiakou area and suggestions on its utilization and protection. *North China Geol.* **2022**, *45*, 44–54.
38. Wang, C.; Zhang, J.; Peng, B.; Wang, J.; Yu, J.; Wu, J. Research on the optimization of land spatial pattern based on "dual evaluation": Taking Anyuan, Jiangxi province as an example. *North China Geol.* **2023**, *46*, 69–78.
39. Yin, Z.; Zhao, L.; Liu, W.; Li, R.; Shao, H.; Peng, C.; Tian, Y. Shrinkage reasons and countermeasures of Moon Lake area in the eastern part of Bashang Plateau, Chengde City. *Hydrogeol. Eng. Geol.* **2020**, *47*, 57–64.
40. Tucker, C.J. Red and photographic infrared linear combinations for monitoring vegetation. *Remote Sens. Environ.* **1979**, *8*, 127–150. [CrossRef]
41. Peng, J.; Liu, Z.; Liu, Y.; Wu, J.; Han, Y. Trend analysis of vegetation dynamics in Qinghai Tibet Plateau using Hurst Exponent. *Ecol. Indic.* **2012**, *14*, 28–39. [CrossRef]
42. Zhao, Z.; Liu, J.; Peng, J.; Li, S.; Wang, Y. Nonlinear features and complexity patterns of vegetation dynamics in the transition zone of North China. *Ecol. Indic.* **2015**, *49*, 237–246. [CrossRef]
43. National Development and Reform Commission, Ministry of Environmental Protection. *Guidelines for Drawing Red Lines for Ecological Protection*; General Office of the Ministry of Environmental Protection: Beijing, China, 2017.
44. Zhu, Z.; Wang, X.; Rao, S.; Zhang, X.; Li, C.; Li, C.; Li, D.; Mu, X. Study of the method on zoning of territorial space ecological protection and rehabilitation and case of Chengde. *Environ. Ecol.* **2020**, *2*, 1–7.
45. Liu, J.; Ma, S.; Gao, J.; Zou, C.; Wang, J.; Liu, Z.; Wang, L. Delimiting the ecological conservation redline at regional scale: A case study of Beijing-Tianjin-Hebei region. *China Environ. Sci.* **2018**, *38*, 2652–2657.
46. Li, J.; Wang, Y.; Yin, S.; Cui, L.; Deng, X. Evaluation of ecological conservation importance and regionalization of dominant ecological function in Zhangjiakou. *Chin. J. Soil Sci.* **2020**, *51*, 280–288.

water

MDPI

Article

Phosphorus Retention in Treatment Wetlands? A Field Experiment Approach: Part 2, Water Quality

Mohamed Z. Moustafa [1,2,*,†] and Wasantha A. M. Lal [1,†]

[1] South Florida Water Management District, 3301 Gun Club Road, West Palm Beach, FL 33406, USA; wlal33463@gmail.com

[2] South Florida Engineering and Consulting, Lake Worth, FL 33460, USA

* Correspondence: zmoustafa@sfec.us; Tel.: +1-561-412-6997

† These authors are retired.

Abstract: In this study, we hypothesized and tested that physical parameters (flow, transport, and water depth) have a significantly greater influence on phosphorus (P) retention in wetlands than biogeochemical factors. Specifically, we evaluated the null hypothesis (H_0), that no significant difference exists between the influence of physical and biogeochemical parameters on phosphorus retention, against the alternative hypothesis (H_1), that physical parameters are more influential. We investigated two large wetlands (stormwater treatment areas, STAs) in south Florida: STA34C2A, which is dominated by emergent aquatic vegetation (EAV), and STA2C3, which is dominated by submerged aquatic vegetation (SAV). Building on Part 1, which mapped spatial flow resistance (K) as a vegetation-type-independent proxy for hydraulic resistance, this study (Part 2) applied a novel high-frequency (hourly) data approach with time-lagged regression modeling to estimate total phosphorus (TP) outflow concentrations. The key variables included inflow TP concentration, vegetation volume, water depth, nominal hydraulic residence time (HRT), hydraulic loading rate (HLR), phosphorus loading rate (PLR), and time lag ("P-spiral"). Multi-linear regression models for each STA identified inflow TP and water depth, a controllable physical parameter, as the most significant predictors of TP outflow, while the hour of day (a temporal proxy) contributed the least. Optimal model performance occurred with lag times of 8 and 9 days, producing R^2 values of 0.5788 (STA34C2A) and 0.5354 (STA2C3). In STA34C2A, high TP retention was linked to shallow water depth, dense EAV, and low K values, indicating high hydraulic resistance and reduced short circuiting. In contrast, lower TP retention in STA2C3 was associated with longer flow paths, sparse SAV, and high K values, suggesting less hydraulic control despite similar nominal HRTs. These results provide empirical support for rejecting the null hypothesis (H_0) in favor of the alternative (H_1): physical parameters, especially water depth, hydraulic resistance, and inflow dynamics, consistently exert a stronger influence on P removal than biogeochemical factors such as PLR. The findings highlight the importance of optimizing flow and depth controls in wetland design and management to enhance phosphorus removal efficiency in large, constructed wetland systems.

Keywords: phosphorus; wetlands; vegetation; residence time; restoration; human activity; protection; management strategies; hydraulic modeling; lag time

1. Introduction

Wetlands serve as important natural and constructed buffers that mitigate phosphorus (P) transport from terrestrial to aquatic systems and are widely recognized for their role in

https://doi.org/10.3390/w17121746

improving water quality by acting as sinks for nutrients, particularly P, which is a leading contributor to eutrophication in aquatic ecosystems [1,2]. Phosphorus, a key nutrient driving eutrophication in freshwater environments, is retained within wetland systems through a range of biogeochemical and physical processes (Figure 1). Biogeochemical mechanisms such as sediment interactions, redox conditions, microbial activity, and vegetation uptake have been widely studied and are commonly recognized as primary drivers of P retention [3,4]. Meanwhile, physical processes, including hydraulic loading rate (HLR), hydraulic residence time (HRT), water depth, and internal flow dynamics, have received comparatively less attention, particularly at the system scale. Wetlands' effectiveness regarding P retention rely on both biogeochemical and physical parameters, operating over various spatial and temporal scales. Understanding the interactions between these factors is crucial for predicting and enhancing P retention.

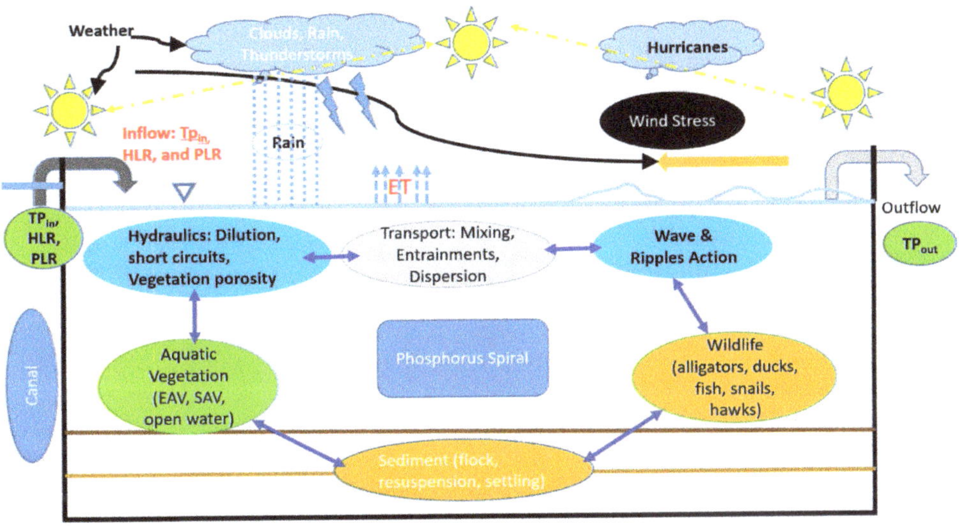

Figure 1. Model schematic of a wetland system depicting major sources and sinks representing biogeochemical and physical processes for phosphorus between inflow and outflow locations.

Previous studies have shown that vegetation can influence both biogeochemical and hydrodynamic processes by slowing water flow, promoting sedimentation, and altering nutrient cycling [5]. Yet, despite this dual role, most investigations have emphasized biogeochemical attributes over the physical mechanisms that directly govern water movement and solute transport. In fact, design guidelines for constructed wetlands often treat physical parameters as fixed, rather than as dynamic and manageable variables. This oversight limits the ability to optimize wetlands for maximum phosphorus retention.

The interactions between vegetation and flow were also investigated in a series of publications [6–8], demonstrating how vegetation resistance and P retention depends on flow type (steady vs. pulsed) and flow characteristic (e.g., laminar, turbulent, or transient).

Despite the recognized importance of biogeochemical processes, physical parameters such as HLR, HRT, flow path, and water depth are also critical to P removal. Yet, those parameters receive disproportionately less attention in the literature. Studies that do explore these variables often do so in isolation or use generalized empirical models that fail to account for their interactions or site-specific behavior [9–11].

In the first phase of this research [12], we demonstrated that wetland vegetation substantially alters flow patterns, creating low-velocity zones favorable for sedimentation and P retention. We introduced the parameter "wetland transmissivity" (K) to quantify vegetation-induced flow resistance, revealing how vegetation density influences residence time and, by extension, P removal performance. These findings emphasized the need to better understand and quantify how physical parameters shape retention outcomes.

We [12] also investigated how wetland vegetation influences internal flow patterns. Using hydrodynamic modeling and field measurements, we demonstrated that plant density/type and species distribution significantly alter residence time distribution and create low-velocity zones conducive to phosphorus sedimentation. This aligns with previous findings that vegetation enhances P retention by increasing particle settling and promoting conditions favorable for P sorption onto sediments [9]. Furthermore, ref. [12] introduced a surrogate parameter, "wetland transmissivity" (K), to represent vegetation resistance to water flow. Vegetation porosity remains independent of both vegetation type and flow regime [12]. High K values correspond to sparse vegetation, or open water, and are associated with lower resistance and a faster flow regime. Conversely, low K values indicate higher plant density and higher vegetation resistance, consistent with slower moving water [12]. These findings emphasized the need to better understand and quantify how physical parameters shape P retention outcomes.

In the current study (Part 2), we build on these findings by systematically evaluating how key physical drivers such as HLR, HRT, water depth, and flow path, affect P retention in a large, constructed wetland. This study utilizes high-resolution, hourly data to investigate the relationships between hydraulic processes and total phosphorus (TP) concentrations at inflow and outflow points, offering a more refined subtle understanding of the physical controls on P removal than traditional low-frequency sampling approaches.

HLR and HRT are especially important because they directly influence the contact time between water and reactive substrates, sedimentation rates, and redox conditions, all of which mediate P transformations [13,14]. Water depth, in turn, affects thermal stratification, vegetation establishment, and the spatial distribution of aerobic versus anaerobic zones, which are crucial for P cycling. However, most wetland design guidelines treat these factors as fixed inputs rather than dynamic, manageable variables.

Wetland size (aspect ratio), morphology, and flow path also impact P retention. Shallow, low-gradient wetlands promote laminar flow, enhancing P sedimentation. Channelization or deep zones may lead to short circuiting, reducing retention. Previous studies [7,9,15] have found that increasing the flow path and contact time between inflow water and wetland substrate increased total P retention. Meanwhile, ref. [7] proposed a diagram representing P retention in terms of flow per wetland's unit width, water depth, and water slope.

This oversight of physical parameters limits the potential to optimize wetland design and operation for P removal. As ref. [5] notes, understanding how these hydrodynamic controls interact with nutrient loads is vital for enhancing treatment performance, particularly in large-scale or agricultural catchment settings.

The second phase of our research (Part 2) aims to bridge this knowledge gap by systematically evaluating the impact of physical drivers such as HLR, HRT, and water depth on P retention in a constructed wetland. We hypothesize that among these variables one or more may emerge as both highly influential and controllable through practical design or operational interventions. To test this hypothesis, we conducted extensive phosphorus concentration monitoring at both the inflow and outflow of a large wetland system over two experimental periods and varying hydrologic conditions.

This data-driven approach offers new insights into the treatment performance of constructed wetlands. The primary focus of the current research is to investigate the relationship between hydraulic processes (i.e., water transport) and phosphorus retention. A clearer understanding of this relationship can inform strategies for optimizing P removal by managing hydraulics. Particularly, how water flows through varying vegetation communities (e.g., EAV, SAV) and how those communities influence key parameters such as HRT, water depth, and subsequently the overall P removal efficiency. Our goal is to identify which physical factors are most strongly associated with P retention and to develop evidence-based recommendations for wetland design and management. By considering this, we aim to provide practitioners and researchers with a framework that balances both biogeochemical and physical considerations, advancing the performance of wetlands as sustainable nutrient removal systems.

In summary, this investigation centers on a fundamental question: can physical parameters, which are directly controllable through wetland design and operational strategies, exert a greater influence on P retention than biogeochemical factors, which are typically less flexible or more difficult to manipulate? If so, this would have major implications for wetland management and design, offering practitioners more precise tools to enhance water treatment performance.

To explore this question, we test the following hypotheses:

- Null Hypothesis (H_0): there is no significant difference between the influence of physical parameters and biogeochemical factors on phosphorus retention in wetlands.
- Alternative Hypothesis (H_1): physical parameters (flow, transport, and water depth) have a significantly greater influence on phosphorus retention in wetlands than biogeochemical factors.

By focusing on the interaction between flow dynamics and wetland structure, including (EAV and SAV), this study seeks to identify which physical variables most strongly correlate with P retention performance. Ultimately, we aim to develop evidence-based recommendations for the design and operation of treatment wetlands that integrate both hydrodynamic and ecological considerations, offering a framework for more effective nutrient management at landscape scales.

2. Materials and Methods

2.1. Experiment Locations

Two experimental sites were selected to represent contrasting vegetative regimes: the first site is dominated by submerged aquatic vegetation (SAV) at STA-2 Cell 3 (STA2C3; Figure 2A), while the second site is dominated by emergent aquatic vegetation (EAV) at STA-3/4 Cell 2A (STA34C2A; Figure 2B). Site selection was also guided by the aspect ratio of the wetland cells, a dimensionless metric describing the proportional relationship between width and length in a rectangular system. The aspect ratios of STA2C3 and STA34C2A were 0.43 and 0.78, respectively, indicating that STA2C3 was significantly more elongated (4600 m vs. 3600 m in length) compared with STA34C2A (Figure 2A,B).

STA2C3 is characterized by two vegetation types (Figure 2A). The top and the western part of Cell 3 is dominated by SAV and open water areas, while the eastern part and the western lower section are dominated by EAV. The section with the EAV is heavily vegetated and may only be accessed via helicopter. The cell interior is intercepted in the longitudinal and transverse directions by remnant agricultural canals. Most of the transverse canals are perpendicular to the southerly flow and dominated by EAV, which affected water flow. During this experiment, flow (HLR), water depths, and P concentrations at inflow and outflow were collected in STA2C3 between 30 August 2013, and 2 December 2013. A hydraulic wave experiment with a pre-determined frequency (Part 1) was conducted between 8 October 2013, and 30 October 2013.

A

Cell	Coverage Type	Estimated Coverage (acre)	Coverage (%)	Cell Area (acre)
Cell 1	Water/SAV	114.51	5.63	2032.56
	EAV	1918.00	94.36	
	Bare Soil	0.04	0.00	
Cell 2	Water/SAV	827.84	34.88	2373.19
	EAV	1544.99	65.10	
	Bare Soil	0.36	0.02	
Cell 3	Water/SAV	1496.01	65.15	2296.33
	EAV	800.08	34.84	
	Bare Soil	0.24	0.01	
Cell 4	Water/SAV	1362.41	70.16	1942.00
	EAV	576.30	29.68	
	Bare Soil	3.29	0.17	

B

Cell	Coverage Type	Estimated Coverage (acre)	Coverage (%)	Cell Area (acre)
CELL 1A	Water/SAV	385.46	12.77	3019.60
	EAV	2634.02	87.23	
	Bare Soil	0.12	0.00	
CELL 1B	Water/SAV	2065.08	59.75	3456.10
	EAV	1389.85	40.21	
	Bare Soil	1.19	0.03	
CELL 2A	Water/SAV	265.94	10.61	2505.55
	EAV	2239.55	89.38	
	Bare Soil	0.05	0.00	
CELL 2B	Water/SAV	1446.37	60.35	2396.45
	EAV	949.79	39.63	
	Bare Soil	0.28	0.01	
CELL 3A	Water/SAV	94.98	3.93	2415.16
	EAV	2320.15	96.07	
	Bare Soil	0.02	0.00	
CELL 3B	Water/SAV	1164.71	55.82	2086.69
	EAV	921.71	44.17	
	Bare Soil	0.26	0.01	

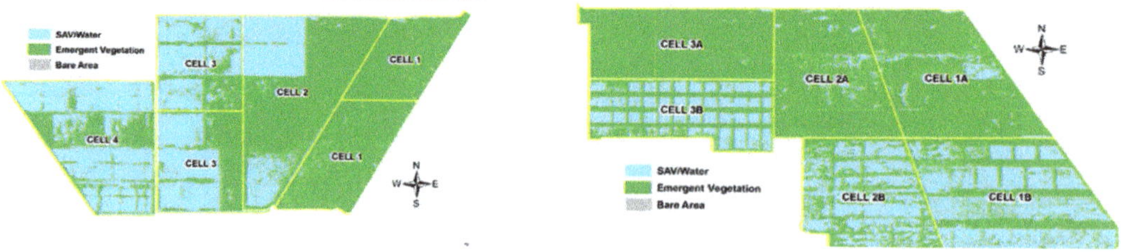

Figure 2. Stormwater treatment area vegetation maps including emergent aquatic vegetation (EAV) represented by light green color and submerged aquatic vegetation (SAV) represented by light blue color. Panel (**A**)—STA2C3, panel (**B**)—Cell34C2A. Percentage distribution of vegetation type for each cell is also included, where Cell 3 of STA2 is covered with >65% SAV and Cell 2A of STA34 is dominated by >96% EAV.

STA34C2A is EAV dominated (Figure 2B). Flow, water depth, and P concentrations at inflow and outflow were collected between 19 May 2014, and 1 February 2015 for the P removal experiment. The hydraulic wave test with a pre-determined frequency was conducted from 4 August 2014 to 24 September 2014. Missing data were excluded from the analysis when there were no discharges at the inflow site in both cells (i.e., G333 and G377).

Water level loggers were deployed in both cells at numerous locations to monitor water surface elevations at 15 min intervals (Figure 1; [8]). Data collected by these loggers were used to determine a surrogate (K) for vegetation resistance [7,8], calculate dynamic water levels (i.e., water depth) inside the cell during the experiments [8], and determine flow type (fast or slow/plug flow type), independent of vegetation type.

2.2. *Definitions of Key Variables*

2.2.1. Hydraulic Loading Rate (HLR)

HLR is calculated using the following equation:

$$HLR = 86,400\frac{\mathrm{Q}}{\mathrm{A}} \tag{1}$$

where Q is surface water inflow ($m^3\ s^{-1}$), A is the effective treatment (wetted perimeter) area of the constructed wetland (m^2), and HLR is the hydraulic loading rate in meter day^{-1}, and 86,400 is the number of seconds in a day.

2.2.2. Water Depth (D)

Water depth is commonly calculated as the average stage between tail (at inflow structure) and head (at outflow structure) water stages, minus the average ground elevation inside the wetland [16–18]. A dynamic water depth was calculated as the average of all observations from all interior stations deployed inside the wetland (e.g., 20 and 23 pressure sensors in both experiments) minus the average ground elevation. The dynamic water depth was calculated on an hourly basis from all water level loggers' measurements. There was almost no significant difference ($\alpha < 0.05$) between the two means from the two different methods, and, therefore, the average water depth for this analysis was calculated using the conventional method [(tail + head)/2 − (average cell ground elevation)].

2.2.3. Nominal Hydraulic Residence Time (HRT_1)

The treatment efficiency of a wetland depends on HRT [19–25]. HRT is the amount of time a parcel of water, for example, introduced through the inflow, travels through the wetland, and leaves at the outflow structures. It is hypothesized that outflow P concentration will decrease as HRT increases, yet it is not a linear relationship. As water can move at different rates and volumes through large, constructed wetlands, particularly in those heavily vegetated STAs (EAV or SAV), the "actual" or true HRT is hard to measure. Hydraulic retention time (HRT) is the average amount of time a compound/nutrient remains in a treatment wetland. In general, a nominal HRT or HRT_1 (in days) is easily calculated using a simple equation [1,26–28]:

$$HRT_1 = \frac{A \cdot D}{Q_{out} \cdot 86,400} \quad (2)$$

where Q_{out} is the surface water outflow in $m^3\ s^{-1}$ (i.e., G334 or G377), A is the constructed wetland's effective treatment area in m^2, and D is the water depth in meters. Nominal HRT_1 is commonly known as HRT and, for most STAs in south Florida, the annual average HRT_1 is approximately 2 weeks (14 days). HRT_2 is calculated by Equation (2), where A is calculated by subtracting the area occupied by vegetation from the total wetland area (wetland's effective treatment area in m^2; Figure 2). Of particular interest is the approach to correct HRT estimates based on errors introduced by ignoring plant volume (i.e., phytomass) and topographic uncertainty in estimating water volume in a heavily vegetated wetland system. Water volume projected by the novel approach, below 2 feet water depth, overestimated water volume by 60 percent, a significant impact on estimating HRT values in a heavily vegetated wetland. HRT also can be determined through conservative tracer studies, which provide a one-time value, a snapshot. Tracer studies are expensive to implement in a large, constructed wetland, designed to determine hydraulic efficiency, and areas of short circuiting [1,26–28].

2.3. Water Quality Sampling Method

Phosphorus concentrations ($\mu g\ L^{-1}$) were measured at sub-daily intervals with two remote phosphorus analyzers (RPA). The RPA combines automated sampling equipment with an automated micro-laboratory and is able to collect a considerable number of TP samples at remote locations for extended periods of time [19]. One RPA was deployed at the center structure of the inflow and outflow of each cell during the experiments. RPA sampling intervals started a few weeks prior to the start of the hydraulic field experiment, at 3-h, then 1.5-h sampling intervals during both field experiments and thereafter for 1 week, and then back to the 3-h sampling intervals. RPA data gaps existed at STA34C2A due to a power failure (lightning strikes). Linear interpolation was used to fill in small data gaps, while HEC-DSSVue (https://www.hec.usace.army.mil/software/hec-dssvue/downloads.

aspx, accessed on 21 February 2025) was used to linearly interpolate those observations into hourly datasets. HEC-DSSVue is a Java-based visual utilities program that allows users to plot, tabulate, edit, and manipulate data in a HEC-DSS database file. The data were combined with discharge inflow ($Q = m^3\ s^{-1}$) and the wetland effective treatment area ($A = m^2$) to calculate phosphorus loading rate ($PLR = g\ m^{-2}\ year^{-1}$):

$$PLR = \frac{TP_{in} \cdot Q \cdot 31,556.952}{A} \tag{3}$$

RPA data for STA2C3 covered the P experiment period from 30 August 2013, to 2 December 2013 (Figure 3); the hydraulic experiment period began on 8 October 2013, and ended on 30 October 2013 (Figure 4). Data collection during the hydraulic wave experiment was hourly as often as possible. All hydraulic and P data collected at the inflow and outflow structures for STA34C2A were reviewed for missing data. The final data analysis period extended from 20 June 2014, to 1 February 2015, and from 4 August 2014, to 24 September 2014, representing P removal and hydraulic sampling experiment periods, respectively (Figures 5 and 6).

Hourly discharges and TP concentrations at inflows (G333 and G377) and outflows (G334 and G378) were used to calculate TP/mass loading (Equation (3)) and hydraulic loading rates (Equation (1)) for further data analysis. Average water depth within Cell 3 and Cell 2A, collected on an hourly basis, were obtained by subtracting average ground elevation (topographic survey in 2010) from the average stage elevations at inflow and outflow structures. All variables used in the analysis were either collected or interpolated (using HEC-DSSVue) to obtain hourly values and were combined into a single database for further analysis.

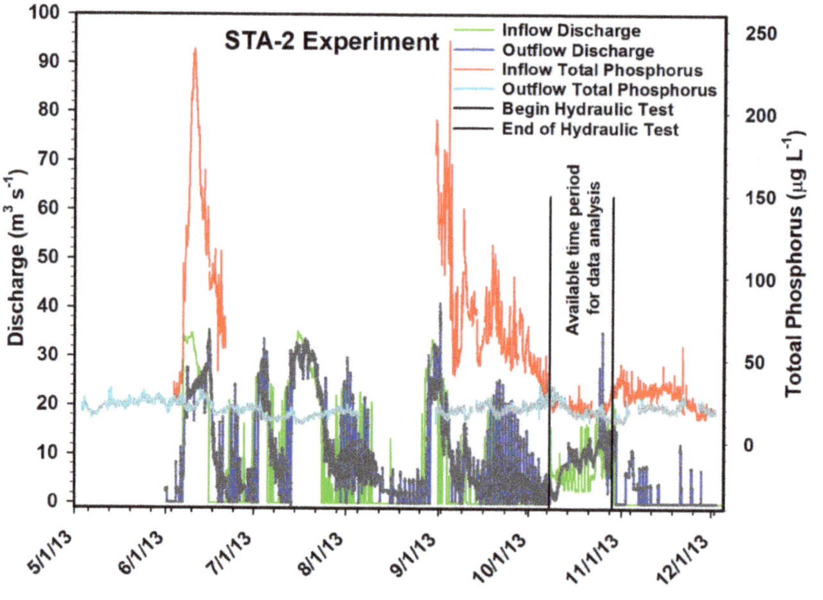

Figure 3. All available field-collected data during current experiment from 1 May 2013, to 1 December 2013. Inflow (solid green line) and outflow (solid blue circles) discharges, and TP concentrations at the inflow (solid red line) and outflow (solid light blue triangle) measured during the P removal experiment in STA2C3 from 1 May 2013, to 1 December 2013. Data gaps shown are due to operation schedule (flood protection), RPA malfunction (due to the breaking of RPA trailer bottom; large gap), and lightning strikes (small gaps, 3 to 4 days). Vertical black lines represent start and end of hydraulic (pulsing) field experiment.

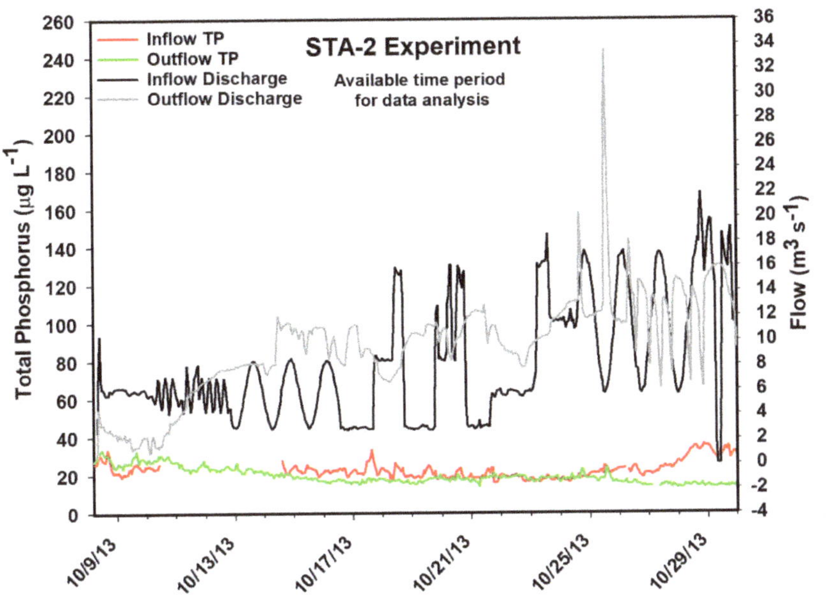

Figure 4. Final hourly database period that contained all variables and no missing data in STA2C3. Database period extended from 30 September 2013, to 31 October 2013. Data before starting date and after ending date contained too many data gaps due to flood control operations and lightning strikes that caused the RPA to fail collecting TP samples. Periods of no inflow discharges (Q = zero) are excluded from HLR and PLR calculations.

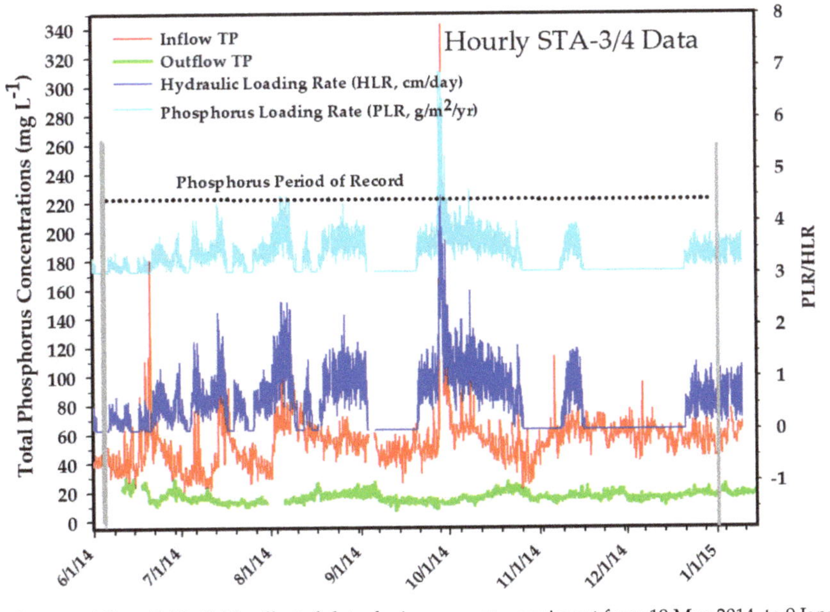

Figure 5. All available field-collected data during current experiment from 19 May 2014, to 9 January 2015. Inflow hydraulic loading rate (solid blue line) and outflow TP loading rate (solid light blue line), and TP concentrations at the inflow (red solid line) and outflow (solid green line) measured during the hydraulic field experiment in STA34C2A for the entire period of recording between 19 May 2014, and 9 January 2015. Data gaps shown are due to lightning strikes and RPA malfunction.

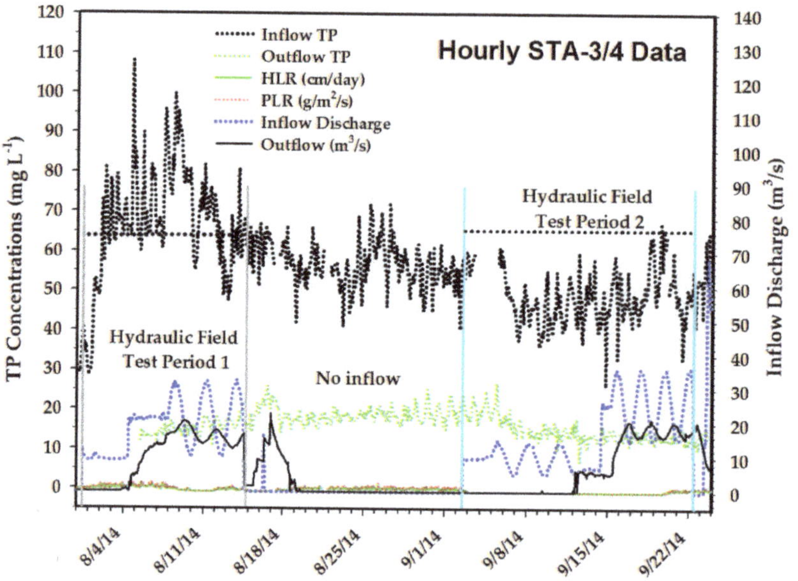

Figure 6. Final database period that contained all variables and contained no missing data in STA34C2A. Database period 1 extended from 31 July 2014, to 14 August 2014, and data period 2 from 2 September 2014, to 22 September 2014. Database period for TP extended from 19 May 2014, to 16 January 2015. Periods of no inflow discharges are excluded from HLR and PLR calculations. Other data monitoring periods are excluded because of too many data gaps due to flood control operations and lightning strikes that caused the RPA to fail collecting TP samples.

2.4. Data Analysis

A combination of descriptive statistics together with graphical techniques were employed to identify seasonality or trends, and describe relationships, if any, in the time series observations. All analyses were performed with the hourly values calculated from the original observations. Statistics were calculated with JMP (Version 12, SAS Institute Inc., Cary, NC, USA) and SigmaPlot (Version 13, SPSS, Inc. Chicago, IL, USA). The level of significance (α) was set at 0.05 for all analyses. Other techniques were used: box and whisker plots were developed and summarized (i.e., top and bottom of box = 75th and 25th percentiles, respectively, midline in box = median; ends of whiskers denote 5th and 95th percentile, respectively, solid black line = mean values). Linear and multi-linear regressions were used to assess the relationship between P concentrations at the outflow sites (G334 and G378), and other key variables such as HLR, PLR, HRT, water depth, and time of day (hour).

3. Results

3.1. Phosphorus Concentration Dynamics During Experimental Periods

A marked discrepancy in TP concentration values was observed between the P removal experiment and the hydraulic field experiment in STA2C3, a trend not evident in STA34C2A (Tables 1 and 2). During the P removal period (30 August–2 December 2013), mean TP concentrations in STA2C3 were 47 ± 0.73 µg L^{-1} at the inflow and 20 ± 0.08 µg L^{-1} at the outflow. In contrast, during the hydraulic experiment period, mean TP concentrations were 23 ± 0.22 µg L^{-1} at the inflow and 19 ± 0.20 µg L^{-1} at the outflow (Table 1).

Table 1. Descriptive statistics of all variables (hourly intervals) used in the analysis for the period 30 August 2013, through 2 December 2013, and during the hydraulic field test (8 October 2013, through 30 October 2013) in STA2C3.

P Removal Sampling Period: 30 August 2013, Through 2 December 2013 (STA2C3)											
Variables	Size	Missing	Mean	Std. Dev	Std. Error	Range	Max	Min	Median	25%	75%
Q_{in} ($m^3\ s^{-1}$)	2265	0	3.53	6.40	0.14	32.45	32.45	0.00	0.00	0.00	5.56
Q_{out} ($m^3\ s^{-1}$)	2265	0	5.41	6.35	0.13	41.73	39.69	−2.04	3.46	0.00	8.93
TP_{in} ($\mu g\ L^{-1}$)	2265	141	47.00	34.00	0.73	216.00	231.00	15.00	34.00	24.00	60.00
TP_{out} ($\mu g\ L^{-1}$)	2265	250	20.00	4.00	0.08	26.00	38.00	12.00	20.00	18.00	23.00
Water Depth (cm)	2265	0	55.40	11.40	0.24	74.90	102.10	27.20	53.00	48.50	60.70
PLR ($g/m^2/yr$)	2265	0	0.77	2.49	0.05	19.35	19.35	0.00	0.00	0.00	0.28
HLR (cm/day)	2265	0	3.39	6.16	0.13	31.21	31.21	0.00	0.00	0.00	5.35
HRT_2 (days)	2265	0	19.90	20.50	0.40	63.0	63.20	0.10	8.50	4.30	41.2
HRT_1 (days)	2265	0	27.90	22.60	0.50	65.8	70.60	4.80	14.80	9.30	50.3
Hydraulic Field Test Period Sampling Period:8 October 2013, through 30 October 2013 (STA2C3)											
Q_{in} ($m^3\ s^{-1}$)	530	0	8.08	4.61	0.20	21.90	21.90	0.00	6.43	4.99	11.42
Q_{out} ($m^3\ s^{-1}$)	530	0	9.51	3.99	0.17	32.75	33.79	1.04	10.09	7.70	11.76
TP_{in} ($\mu g\ L^{-1}$)	530	104	22.78	4.59	0.22	20.50	36.50	16.00	21.88	19.34	24.41
TP_{out} ($\mu g\ L^{-1}$)	530	4	19.29	4.51	0.20	20.75	33.75	13.00	18.00	16.25	22.00
Water Depth (cm)	530	0	59.40	7.28	0.32	33.83	74.98	41.158	60.05	57.00	63.09
PLR ($g/m^2/yr$)	530	104	0.69	0.48	0.02	2.51	2.51	0.00	0.56	0.36	0.93
HLR (cm/day)	530	0	7.77	4.43	0.19	21.06	21.06	0.00	6.18	4.80	10.98
HRT_2 (days)	530	0	4.74	4.62	0.20	61.88	63.17	1.29	4.04	3.50	4.29
HRT_1 (days)	530	0	13.24	18.00	0.78	64.50	69.50	5.00	6.67	5.92	9.01

Table 2. Descriptive statistics of all variables (hourly intervals) used in the analysis for the P removal sampling period 19 May 2014, through 1 February 2015, and during the hydraulic field experiment (4 August 2014, through 24 September 2014) in STA34C2A.

P Removal Sampling Period: 19 May 2014, Through 1 February 2015 (STA34C2A)											
Variables	Size	Missing	Mean	Std. Dev	Std. Error	Range	Max	Min	Me-dian	25%	75%
Q_{in} ($m^3\ s^{-1}$)	6196	0	4.96	9.36	0.12	68.41	68.41	0.00	0.00	0.00	6.23
Q_{out} ($m^3\ s^{-1}$)	6196	0	3.82	6.21	0.08	28.31	28.31	0.00	0.25	0.00	6.18
TP_{in} ($\mu g\ L^{-1}$)	5645	142	55.00	19.00	0.26	324.00	343.00	19.00	54.00	45.00	62.00
TP_{out} ($\mu g\ L^{-1}$)	5795	853	17.00	3.00	0.05	23.00	30.00	7.00	16.00	14.00	19.00
Water Depth (cm)	6196	0	56.08	14.32	0.18	60.31	98.27	37.96	50.23	45.78	62.22
PLR ($g/m^2/yr$)	5129	108	1.23	2.91	0.04	38.55	38.55	0.00	0.00	0.00	1.34
HLR (cm/day)	6196	0	4.17	7.86	0.10	57.45	57.45	0.00	0.00	0.00	5.24
HRT_2 (days)	6196	0	18.10	16.20	0.20	66.70	66.80	0.10	13.8	5.50	24.60
HRT_1 (days)	6196	0	22.50	17.20	0.20	74.30	74.30	0.10	18.1	9.30	28.80
Hydraulic Sampling Period: 4 August 2014, through 24 September 2014 (STA34C2A)											
Q_{in} ($m^3\ s^{-1}$)	1248	0	11.92	12.58	0.36	68.41	68.41	0.00	9.68	0.00	21.21
Q_{out} ($m^3\ s^{-1}$)	1248	0	7.09	7.92	0.22	28.31	28.31	0.00	3.47	0.00	15.28
TP_{in} ($\mu g\ L^{-1}$)	1248	77	57.00	11.00	0.33	82.00	108.00	26.00	55.00	49.00	62.00
TP_{out} ($\mu g\ L^{-1}$)	1248	40	16.00	3.00	0.09	20.00	27.00	7.00	16.00	14.00	18.00
Water Depth (cm)	1248	0	67.33	15.83	0.45	54.24	96.58	42.34	68.03	51.85	82.42
PLR ($g/m^2/yr$)	1248	77	2.22	2.46	0.07	13.62	13.62	0.00	1.51	0.00	4.22
HLR (cm/day)	1248	0	10.01	10.56	0.30	57.45	57.45	0.00	8.13	0.00	17.82
HRT_2 (days)	1248	0	11.00	9.50	0.30	28.00	29.60	1.50	6.50	2.50	19.20
HRT_1 (days)	1248	0	15.70	10.30	0.30	29.90	34.50	4.60	12.80	5.80	24.70

Similarly, STA34C2A exhibited consistent P concentrations across both experimental periods. During the P removal period, mean TP values were 55 ± 0.26 μg L^{-1} at the inflow and 17 ± 0.05 μg L^{-1} at the outflow. These values closely align with those observed during the hydraulic period, where TP concentrations averaged 57 ± 0.33 μg L^{-1} and 16 ± 0.09 μg L^{-1} at the inflow and outflow, respectively (Table 2).

The hourly TP concentration measurements at both inflow and outflow locations revealed a consistent diel (24 h) cycle during both experimental phases (Figure 7). TP concentrations increased gradually beginning around 07:00, peaking between 12:00 and 13:00, and declining steadily to a minimum around midnight (Figure 7B). These diel trends are consistent with previous findings on nutrient cycling in wetland systems [20].

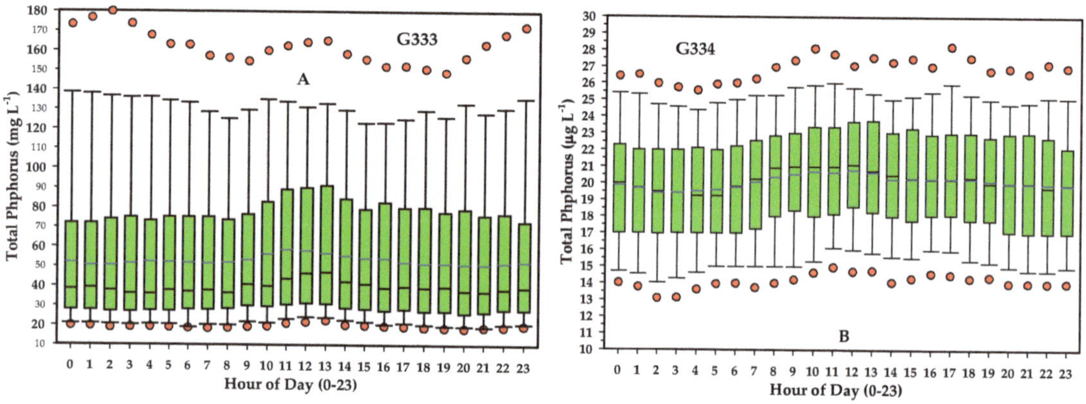

Figure 7. Panel (**A**) and (**B**) box and whisker plots represent inflow (G333) and outflow (G334) TP concentrations during the P removal sampling period in STA2C3. Boxes depict all observed concentrations during the time of day hour from 00 through to hour 23 (midnight). Solid red circles represent the 5th and 95th percentiles, while whiskers represent the 10th and 90th percentiles. Daily trend is more pronounced at both inflow and outflow with peak TP concentrations at midday (hour = 11) and minimum TP concentrations near midnight (hour = 23).

The statistical analysis confirmed significant differences in TP concentrations between diurnal (daytime) and nocturnal (nighttime) periods ($p = 0.05$), with this pattern being more pronounced at the outflow sites (Figure 7B). Paired *t*-tests for STA2C3 further indicated statistically significant differences between the concentrations measured at 10:00 and 23:00 at both inflow and outflow locations ($\alpha = 0.02$). In contrast, for STA34C2A, this time-based difference was significant only at the outflow site (G378; $\alpha = 0.025$), with no corresponding significance observed at the inflow (G377) (Figure 8A,B).

Normality testing using the Shapiro–Wilk test indicated that TP concentration data at inflow sites in both STA2C3 and STA34C2A deviated from a normal distribution, suggesting a stochastic component in P dynamics. For the STA34C2A outflow (G378), the difference in median TP values between 10:00 and 23:00 was not statistically significant ($p = 0.165$), indicating that observed variations may be attributed to random sampling variability. Conversely, the inflow site (G377) exhibited a statistically significant difference in median TP between the same time points ($p = 0.002$), pointing to a non-random, time-dependent fluctuation in TP levels.

The findings of this study highlight the critical role of physical parameters, particularly hydraulic retention time (HRT), water depth, and flow path geometry, in governing P retention in constructed wetlands. These physical factors often surpass biogeochemical processes in determining the efficiency of P removal. For instance, a study by [21] demonstrated that water residence time is a primary factor controlling P retention in small, constructed wetlands treating agricultural drainage water. Their research indicated that during high discharge events, shorter residence times led to reduced P retention, emphasizing the importance of maintaining adequate HRT for effective nutrient removal. Similarly, ref. [22] found that sediment deposition and accumulation, influenced by hydraulic load and flow dynamics, significantly affect P retention in small, constructed wetlands. Their study highlighted that high hydraulic loads can lead to increased resuspension and reduced sediment accumulation, thereby diminishing the wetland's capacity to retain phosphorus. Moreover, a meta-analysis revealed that wetland hydrology and morphometry, including flow regulation and wetland size, are key controls of phosphorus retention [23]. The study emphasized that regulated flow regimes and appropriate wetland sizing can enhance P retention by promoting settling and uptake processes. Meanwhile, ref. [24] demonstrated that the effects of both water depth and wetland area, and also wetland water volume, may be important for N removal.

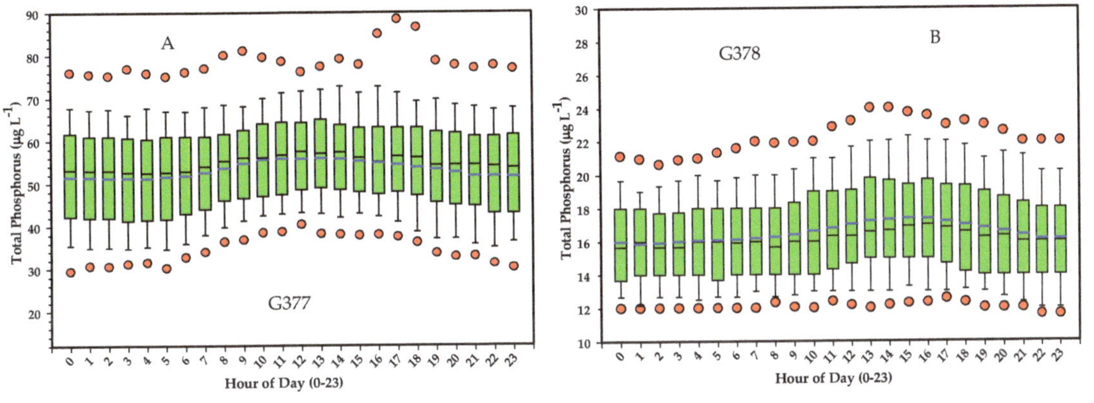

Figure 8. Panel (**A**) and (**B**) box and whisker plots of inflow and outflow TP concentrations at G377 and G378, respectively during the P removal sampling period in STA34C2A. Boxes depict all observed concentrations during the time of day hour from 00 through to hour 23 (midnight). Solid red circles represent the 5th and 95th percentiles, while whiskers represent the 10th and 90th percentiles. Daily trend is more pronounced at both inflow and outflow with peak TP concentrations at midday (hour = 11) and minimum TP concentrations near midnight (hour = 23).

These studies collectively underscore the importance of considering physical design and operational parameters in the development and management of constructed wetlands for P removal. By optimizing factors such as HRT, water depth, and flow path geometry, it is possible to enhance the efficiency of P retention, thereby improving water quality outcomes in agricultural and wastewater treatment contexts.

3.2. Hourly Regression Results

Simple linear regression analyses did not yield statistically significant relationships between outflow TP concentrations and individual predictor variables (e.g., PLR) for either STA2C3 or STA34C2A. This outcome was consistent across both experimental periods (i.e., phosphorus removal and hydraulic field studies) and for both hourly and daily average datasets. In contrast, multiple linear regression (MLR) models provided improved explanatory power compared with univariate models (Figure 9).

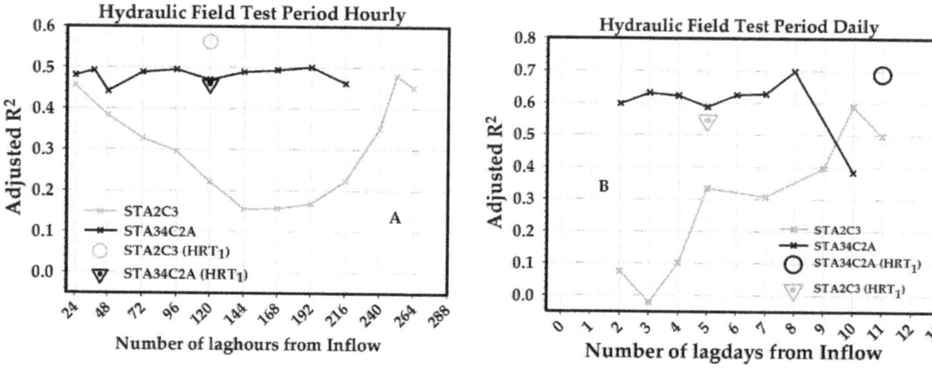

Figure 9. Multi-regression analysis results using all available hourly (panel (**A**)) and daily (panel (**B**)) data (flow and concentrations) during the hydraulic field test period between 8 October 2013, and 30 October 2013, and from 4 August 2014, to 24 September 2014, in STA2C3 (gray line) and STA34C2A (black line), respectively. Three regression model results are presented: first, one with PLR (black solid line) and, second, one without PLR (solid gray line), while the third model added the hour of day to the independent list of variables for STA2C3 (gray circle) and STA34C2A (black triangle), respectively.

In the MLR analysis, a time of day variable ("Hour", ranging from 00 to 23) was incorporated as a predictor to account for diel variation in TP concentrations. The regression model was consistently applied to both stormwater treatment areas (STAs) using the following formulations, where Equations (4) and (5) are used for daily-averaged and hourly data:

$$TP_{out} = PLR + TP_{in} + \text{water depth} + HLR + HRT \quad (4)$$

$$TP_{out} = Hour + TP_{in} + \text{water depth} + PLR + HLR + HRT \quad (5)$$

To evaluate transport time dynamics, a time lag variable (ranging from 0 to 240 h or 0–10 days) was introduced to Equation (5). This lag represents the estimated travel time for P from inflow to outflow, encompassing HRT, biogeochemical spiraling, and short circuiting within the wetland system. When the lag variable was included, HRT was excluded from the regression model to avoid multicollinearity.

During the hydraulic sampling period, the MLR model incorporating a 0-h lag produced adjusted R^2 values of 0.5616 (0.5834 with HRT_1) for STA2C3 and 0.4558 (0.5554 with HRT_1) for STA34C2A (Figure 9A). This indicates an improved model fit when nominal HRT (HRT_1) was used compared with the empirically adjusted HRT_2.

In STA2C3, the most influential predictors were water depth, hydraulic loading rate (HLR), PLR, inflow TP concentration, and HRT_2 (all $p << 0.01$), with the hour of day variable being the least significant. Substituting HRT_1 for HRT_2 yielded a similar ranking, with water depth remaining the most influential variable and hour, again, being the least significant.

Conversely, in STA34C2A, inflow TP concentration emerged as the most significant predictor, followed by HRT_2, water depth, PLR, and hour ($p << 0.01$), with HLR being the least significant. When HRT_1 was used in place of HRT_2, the order shifted slightly: HRT_1 > water depth > inflow TP > PLR > hour ($p < 0.01$), with HLR remaining the least significant.

When the lag variable replaced HRT in the model, the highest adjusted R^2 values indicated optimal lags of 216 h (9 days) for STA2C3 and 192 h (8 days) for STA34C2A during the hydraulic period (Figure 9A). These values are within range but slightly lower than the nominal HRT estimates derived from field data, which were 16 days (384 h) for

STA2C3 and 13 days (312 h) for STA34C2A using HRT_1, and 11 days (264 h) and 5 days (120 h), respectively, using HRT_2 (Figure 9B).

Notably, the inclusion or exclusion of the hour of day variable did not significantly affect the adjusted R^2 values in either STA, suggesting that diel variability, while present, contributes minimally to the overall variability in outflow TP concentrations at this scale.

Finally, the higher adjusted R^2 values obtained using HRT_1 rather than HRT_2 suggest potential inaccuracies in the empirical adjustments applied to derive HRT_2, which accounts for vegetative volume and topographic irregularities. These discrepancies highlight the importance of accurate site-specific hydrological characterization for the predictive modeling of phosphorus dynamics in treatment wetlands.

The multiple-linear regression model results improved using HRT_1 (i.e., nominal HRT) compared with HRT_2 for STA2C3 and STA34C2A, respectively, during the hydraulic sampling period. The noted increase in adjusted R^2 when HRT_1 was included in the place of HRT_2 suggests that the error model used to account for the volume of vegetation and topographic irregularities in the study site to calculate HRT_2 is inaccurate (Figure 9B).

3.3. Daily Regression Results

During the hydraulic sampling period (8–30 October 2013), the daily mean TP concentrations in STA2C3 were 23 ± 1.28 μg L^{-1} at the inflow and 19 ± 4.43 μg L^{-1} at the outflow (Table 3). In STA34C2A, the corresponding values were 56 ± 1.37 μg L^{-1} at the inflow and 16 ± 0.38 μg L^{-1} at the outflow (Table 3). Although inflow TP concentrations exhibited greater variability in STA2C3 compared with STA34C2A (nearly twice the range), the mean and median TP values within each STA were nearly identical, indicating a relatively symmetrical distribution of daily TP measurements.

Table 3. Descriptive statistics of all variables (daily intervals) used in the analysis during the hydraulic field experiment (8 October 2013, through 30 October 2013) in STA2C3 for the period 5 August 2014, through 24 September 2014 (for STA34C2A).

Variables (STA2C3)	Size	Missing	Mean	Std. Dev.	Std. Error	Range	Max	Min	Median	25%	75%
TP_{in}	23	4	23.60	5.59	1.28	20.88	38.50	17.63	22.25	19.63	25.25
TP_{out}	23	0	18.99	4.43	0.92	14.25	27.88	13.63	17.38	16.13	21.25
Water Depth (cm)	23	0	60.50	8.25	1.72	34.70	77.10	42.40	60.70	57.30	63.70
PLR	23	4	0.69	0.53	0.12	2.32	2.32	0.00	0.59	0.37	0.76
HLR	23	0	7.20	4.51	0.94	17.53	17.53	0.00	5.78	4.06	9.00
HRT_1	23	0	15.60	21.33	4.45	65.13	70.38	5.25	6.71	5.83	9.79
Variables (STA34C2A)	**Size**	**Missing**	**Mean**	**Std. Dev**	**Std. Error**	**Range**	**Max**	**Min**	**Median**	**25%**	**75%**
TP_{in} (μg L^{-1})	52	1	56.63	9.80	1.37	44.76	87.54	42.78	55.58	49.63	60.85
TP_{out} (μg L^{-1})	51	0	16.28	2.70	0.38	10.64	22.79	12.15	16.07	13.78	18.64
Water Depth (cm)	52	0	67.33	15.88	2.20	52.15	95.07	42.91	67.59	51.05	82.20
PLR (g/m^2/yr)	52	1	2.17	2.15	0.30	7.91	7.91	0.00	1.61	0.00	4.30
HLR (cm/day)	52	0	10.01	9.32	1.29	33.91	33.91	0.00	8.89	0.00	17.91
HRT_1 (days)	52	0	15.73	10.42	1.45	29.55	34.43	4.88	12.79	5.79	25.00

Multiple linear regression (MLR) models incorporating lag times (i.e., from 1 to 10 days) were applied to assess the delayed response of outflow TP concentrations (TP_{out}) to changes in inflow concentrations (TP_{in}), alongside other key hydraulic and biogeochemical variables (Figure 9B). For STA2C3, the strongest model performance was observed with a 10-day lag, yielding an adjusted R^2 of 0.5889. In this model, TP_{in} was the most influential predictor ($p = 0.056$), followed by hydraulic loading rate (HLR), phosphorus loading rate (PLR), and water depth.

A parallel analysis using hydraulic residence time (HRT) instead of lag days showed that models incorporating nominal HRT (HRT_1) outperformed those using empirically adjusted HRT (HRT_2). Specifically, a model with a zero-day lag and HRT_1 yielded an adjusted R^2 of 0.5562, compared with 0.5440 when HRT_2 was used. Field-based estimates of mean HRT_1 and HRT_2 for STA2C3 during this period were 5 ± 0.20 and 13 ± 0.78 days, respectively (Figure 9B), indicating a potential overestimation by HRT_2 due to the correction factors for vegetation volume and topographic variability.

In STA34C2A, similar trends were observed. The highest adjusted R^2 values occurred with TP_{in} lagged by 8 to 10 days, consistent with the results from the hourly regression analysis (Figure 9B). The strongest model performance was observed at a 10-day lag, while all other lag durations (1–12 days) yielded considerably lower correlations. When comparing HRT metrics, the model using HRT_1 achieved an adjusted R^2 of 0.6900, whereas the model with HRT_2 yielded 0.5727. The mean field-based estimates of HRT_1 and HRT_2 for STA34C2A during the hydraulic sampling period were 11 ± 0.30 and 16 ± 0.30 days, respectively (Figure 9B).

Variable importance rankings, based on significance (p-values), consistently identified TP_{in} as the most significant predictor of TP_{out} in STA34C2A ($p << 0.01$), followed by water depth and HLR ($p << 0.01$), and PLR ($p < 0.01$). These findings reinforce the utility of time-lagged inflow TP concentrations and residence time metrics in predicting phosphorus export dynamics at the outflow.

4. Discussion

4.1. Key Variables Effect

Over the past several decades, extensive research has been conducted on wetlands, with a significant focus on their application in domestic wastewater treatment [25,26] and nutrient reduction, particularly P removal [7,24–30]. Some of these studies have also examined the key variables influencing wetland performance, regardless of whether the primary target was P or N removal. In the past decade alone, the construction of large-scale wetlands (exceeding 40 hectares) has become more common, particularly for treating agricultural and urban runoff [7,29]. These large, engineered wetlands introduce new complexities in estimating standard operational parameters, such as hydraulic loading rate (HLR) and hydraulic retention time (HRT), which are essential for predicting phosphorus removal efficiency. The integration of advanced design elements, such as multiple large cells and controlled inflow and outflow structures, has further complicated the management of these systems for optimal phosphorus removal. For instance, changes in cell size directly influence HLR estimates [1]. These developments prompt a critical question: how can the design and operation of large, engineered wetlands be optimized to enhance P removal performance?

A summary of a recent discussion concerning some of these challenges can be found in [28]. The study [28] concluded that the large-sized wetlands exhibit a more restricted performance spectrum compared with the broader group of wetlands of all sizes, and that seasonality typically has a minimal effect on steady flow systems. This review, however, excluded the impact of P monitoring frequency and hydrology on P retention in those large-sized STAs and wetlands.

The focus of our research was mainly driven by testing and identifying the dominant role in P retention in wetlands; is it biogeochemical processes or physical parameters (transport, flow, and wave) and the monitoring frequency of those parameters that impacts P performance in those systems? Part 2 of our study specifically examined P monitoring frequency in large, constructed wetlands. We employed an unprecedented hourly monitoring approach in two large treatment cells: STA34C2A, characterized by a short travel distance, and STA2C3, with a long travel distance. This high-frequency data allowed us to determine key variables and compare P retention performance based on flow types and sampling frequency.

Field experiments and data collection are essential for understanding cause-and-effect relationships in wetland nutrient removal. Our field experiments aimed to identify which variable or combination of variables can predict TP concentrations at the outflow, thereby improving the management of treatment wetlands for P retention. We also explored how altering the sampling frequency (hourly vs. daily) would affect the influence of these variables on our ability to predict outflow TP concentrations. The data collected provided valuable insights into the internal dynamics of these systems under various operating conditions and existing vegetation.

Interestingly, simple linear regression analysis did not reveal a significant correlation between the outflow TP concentrations and phosphorus loading rate (PLR) or other individual key variables in either STA, regardless of the sampling period or frequency. However, multiple linear regression model results showed a better fit. This analysis introduced two new variables: hour (representing the time of day) and lag (representing the combined effect of hydraulic retention time (HRT), phosphorus spiral, and short circuiting). Model predictions were used to evaluate methods for estimating HRT. For example, the multiple regression model suggested that HRT_2, which accounts for vegetation volume, evapotranspiration (ET), and topographic irregularity, yielded less favorable results than nominal HRT_1, indicating a need for further research on HRT calculation in large, heavily vegetated constructed wetlands.

The multi-regression model results indicated that water depth and inflow TP (TP_{in}) were the most significant factors explaining the variability in outflow TP (TP_{out}) in STA2C3 and STA34C2A, respectively, during the hydraulic sampling period, irrespective of sampling frequency. Water depth consistently influenced TP_{in}. While these findings align with previous research on wetland databases, the ranking of other key variables differed between the two STAs. For STA2C3, water depth was dominant, followed by hydraulic loading rate HLR, PLR, and HRT_2, with hour being the least influential. In STA34C2A, TP_{in} was the most dominant, followed by HRT_2, water depth, PLR, and hour, with HLR being the least influential. HRT_2 generally improved the model's ability to predict TP_{out} compared with HRT_1, highlighting its importance. The higher statistical significance for STA34C2A further underscores the need for improved HRT calculation methods in large, constructed wetlands.

Our findings reveal both similarities and differences between the two STAs, providing insights into cause-and-effect relationships. For instance, lag time and mean water depth were similar in both STAs. However, they differed in which key variable most strongly predicted TP_{out}. Despite similar water depths, TP_{in} was the primary predictor in STA34C2A, while water depth was the primary predictor in STA2C3. Both STAs also had different aspect ratios, with STA2C3 being considerably longer. Although the longer length of STA2C3 suggests more contact time for P removal, STA34C2A exhibited better P performance, achieving the target P concentration ($\leq$13 μg L^{-1}) at the outflow more frequently.

We hypothesize that the superior performance of STA34C2A is due to a more plug-flow-like condition (physical parameter), where water moves through the dense emergent aquatic vegetation (EAV) and reaches the outflow structure uniformly with minimal short circuiting. Contour maps show equal arrival times across the width of STA34C2A. In contrast, STA2C3, dominated by submerged aquatic vegetation (SAV) and open water areas, is more susceptible to short circuiting, with wildlife creating preferential flow paths

(Video S1). Although both STAs had a similar HRT, the dense EAV and shorter travel distance with low transmissivity (*K*) in STA34C2A likely contributed to higher TP removal compared with STA2C3, which has less dense vegetation and higher transmissivity [7].

The available literature concluded that plug-flow conditions enhance P removal [1]. The presence of SAV and open water in STA2C3, leading to braided channels and less resistance to flow, likely resulted in faster water movement along certain paths despite the longer overall distance. This fast movement in STA2C3 resulted in a similar lag time to the slower moving water in the shorter, but densely vegetated, STA34C2A. The contrasting characteristics of the two STAs highlight the importance of vegetation type and flow patterns in influencing P removal efficiency, with the shorter distance in STA34C2A emphasizing the influence of inflow TP concentration and the longer distance in STA2C3 emphasizing the role of water depth.

4.1.1. Hydraulic Residence Time (HRT)

Water flow through wetlands is influenced by resident vegetation, making it challenging to estimate HRT using a simple mathematical expression [7]. Large, heavily vegetated wetlands with diverse vegetation types (EAV, SAV, and open water) create a complex environment for determining hydraulic retention time (HRT), a key variable in predicting TP_{out} outflow concentrations and P removal performance. HRT is one of the key variables to predict TP_{out} and its influence on P removal performance is a function of inflow discharge and vegetation density and resistance. Yet, the accurate prediction of TP_{out} is desired to understand the factors that affect TP performance.

HRT in small and large wetlands is influenced by inflow discharge, plant density, and vegetation resistance. The time lag introduced in our model (Equation (5), Figure 8) represents P residence time, which is a product of hydraulic residence time, P spiral, and short circuiting in a wetland. Yet, the presence of three different water volumes in a wetland [1,28,29,31,32] also impacted HRT, and identified three zones, actively flowing main channel, temporary storage zone, and isolated/dead zone, where the first and second exchange water and constituents, hence impact HRT and P retention. Wind also can play a key role in enhancing mixing and water exchange between the different zones [33].

In large, constructed wetlands, there is a considerable delay between input and output, due to their size. We introduced a delay component as a variable to examine the system (i.e., lag in days), calculating it using data association techniques. The cross-correlation technique assumes stochastic input–output data and calculates the correlation between the two sets. The amount of the delay depends on the operating rules of those wetlands and STAs. We defined this delay factor as the HRT_p for TP_{out} (driven by a combination of the hydraulics, short circuiting, and the P spiral in a wetland, all of which are representative of physical parameters).

The results of the multi-regression model revealed an average of 9 and 10 days of HRT_P (HRT_1 and HRT_2), based on a data association method for STA34C2A and STA2C3, respectively (Figure 9A). The daily results of the multi-linear regression (Figure 9B) confirmed those results (HRT_P = 8 and 8 days). A dye release study also estimated HRT to be 7–10 days in STA-2C3 [22], agreeing with our regression model results. Based on the multi-linear regression model results using lags, the average HRT during the hydraulic period of recording was 10 days. HRT values calculated by the regression model agree with the dye release experiment [22]. This suggests that lag (delay), as calculated using data association techniques, is a new way to estimate HRT_P in large, constructed wetlands.

The observed HRT in STA2C3 is influenced by several factors (Figure 1), including aspect ratio, SAV growth and expansion, and wildlife interactions. SAV provides food and habitats for waterfowl and fish [34], causing changes in SAV presence and affecting

P retention. In contrast, heavily vegetated wetlands with EAV are less prone to wildlife impacts. Water moved slowly through high-density EAV in STA34C2A, reaching the outflow site at the same time as fast moving water in STA2C3, which traveled twice the distance. This may have led to the observed difference in P performance between the two wetlands.

Interaction among all those factors is the cause for the observed P retention, which is less effective in STA2C3 compared with STA34C2A, as evident in the mean and median TP_{out} values (Tables 1 and 2). Numerous interactions in STA2C3, such as wildlife interactions with SAV, SAV interactions with flow (short circuiting), wind drag on water, open water areas, vegetation, etc., compared with STA34C2A impacted the P retention results. Because the results of those interactions are a daily occurrence, in addition to HLR, PLR, water depth, and atmospheric deposition, it caused wide fluctuations in the estimated HRT values. Consequently, wide HRT fluctuations are evident in the STA2C3 parameter values when compared with mean values during the P removal and hydraulic sampling periods (Table 1). The marked differences in key parameter values between the two sampling intervals are only observed in STA2C3, not in STA34C2A (Tables 1 and 2). On the contrary, heavily vegetated wetlands with EAV resulted in slow moving water in STA34C2A (low K values), while fast moving water in STA2C3 (high K values) led to the observed and less significant P performance compared with STA34C2A. A longer distance allows for more contact time with resident biota that should lead to high P performance [1,13,24].

The outcomes of our analysis underscore the critical role of considering the interplay of physical and hydrodynamic factors within expansively constructed wetlands. The significant variations observed in key parameter values between sampling periods at STA2C3, contrasting with the relative consistency at STA34C2A, indicate that emergent aquatic vegetation (EAV) wetlands may offer favorable conditions and foster a more stable environment concerning P retention. We recommend further investigation to elucidate the intricate relationships among the variables that influence P retention in these systems.

4.1.2. Sampling Frequency

Exceedance probability plots are another tool to compare TP_{out} performance (Figure 10). STA34C2A clearly achieved TP_{out} values less than 13 μg L^{-1} for almost 18% of the experiment time, regardless of the time period analyzed (hourly or daily) during both hydraulic P removal experiment periods (Figure 10A). In contrast, STA2C3 achieved TP_{out} values of 13 μg L^{-1} for less than 3% of the experiment time (Figure 10B). A better performance in TP_{out} between 13 and 23 μg L^{-1} for STA2C3 compared with STA34C2A was observed, which may be related to the wave frequency set during the experiment (Figure 10). A more recent study demonstrated higher dispersion (hydrodynamics and physical parameter) values with generated hydraulic waves [8] compared with steady state flow [35]. Dispersion is a key factor in spreading pollutants and nutrients in water bodies such as lakes, wetlands, and rivers/streams. We speculate that the observed high performance in STA2C3 (TP_{out} between 13 and 23 μg L^{-1}) during these experiments may have been caused by wave-generated high-dispersion values (Figure 10B).

The high phosphorus (P) removal performance observed in STA34C2A can primarily be attributed to the combination of its aspect ratio and vegetation density index (as detailed in Part 1), alongside its plug flow-type characteristics resulting from low hydraulic conductivity (K) values and high vegetation resistance (also discussed in Part 1). This is substantiated by the considerable number of observations indicating low P concentrations at the outflow (Figure 10A).

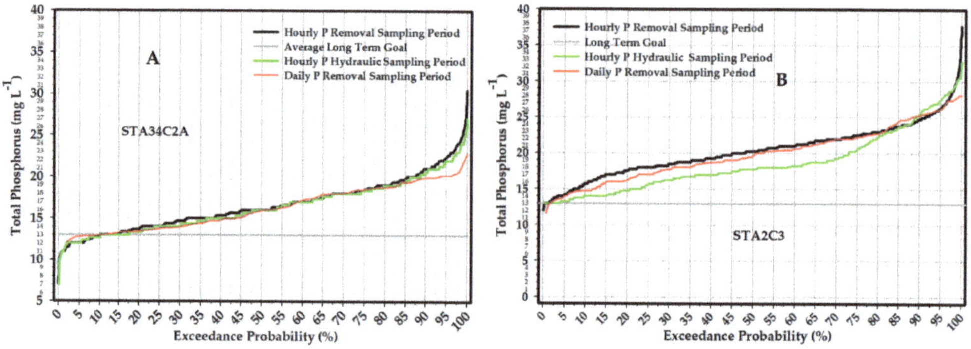

Figure 10. Exceedance total phosphorus concentrations at STA34C2A (panel (**A**)) and STA2C3 (panel (**B**)) outflow structure (TP_{out}) during the entire sampling period (solid black line) and during P removal (solid red line) and hydraulic sampling experiment (solid green line) period, respectively.

Despite STA2C3 having an aspect ratio twice as large as STA34C2A, the latter still achieved high P removal. This success is likely due to its aspect ratio in conjunction with the resident vegetation, specifically emergent aquatic vegetation (EAV) as opposed to submerged aquatic vegetation (SAV). The dense EAV type contributes to significant vegetation resistance, as evidenced by the low observed *K* values and vegetation index (Part 1), thereby fostering a plug flow regime.

Although median total phosphorus (TP) concentration values were comparable for the two time periods across both STAs (Tables 1 and 2), observed TP concentrations were consistently lower in STA34C2A relative to STA2C3, suggesting superior P retention. In light of these variations in P retention, the aspect ratio (length/width) and vegetation type (EAV, SAV) warrant careful consideration in the design of future wetlands and the management of large, constructed wetlands.

Figures 8–10 of Part 1 [12] show contour maps of arrival time at locations inside the cells. Porosity (i.e., plant density index) clearly demonstrates that vegetation resistance affects flow type and, consequently, TP removal performance. The observed change in TP concentrations in a treatment cell is likely to be lower at the outflow compared with the inflow and is determined, among other factors, by vegetation resistance or porosity (hydraulics). High and low porosity values [12] resulted in the observed high and low P retentions in STA34C2A and STA2C3, respectively.

4.1.3. Time of Sampling

RPA hourly sampling during the experiment revealed an interesting phenomenon, in which P concentration changed as a function of "hour of day" or sampling time (Figures 7 and 8). This trend is clearly demonstrated at the outflow compared with the inflow sites and is also statistically significant among day and night hours. For example, for the *t*-test results at G377 (10:00 AM and midnight hour 23), the difference in the median values between the two groups is greater than would be expected by chance, i.e., there is a statistically significant difference ($p = 0.002$). The difference in the median values between the two groups (07:00 AM and midday/noon hour 12:00) at G377 is statistically significant ($p = 0.002$). However, the difference in the median values between the two groups at G377 (07:00 AM and midnight hour 23) is not statistically significant ($p = 0.532$). Meanwhile, the difference in the median values between the two groups (10:00 AM and midnight hour 23) at G378 is not statistically significant ($p = 0.165$). Also, the difference in the median values between the two groups (07:00 AM and midnight hour 23) at G378 is not statistically significant ($p = 0.744$). Yet, the difference in the median values between the two groups

(07:00 AM and midday/noon hour 12:00) at G378 is greater than would be expected by chance, i.e., statistically significant ($p = 0.019$).

4.2. Hypothesis Testing and Interpretation of Results

This study was designed to evaluate the relative influence of physical and biogeochemical factors on P retention in large, constructed wetlands, testing the following hypotheses: the null hypothesis (H_0) postulates that there is no significant difference between the influence of physical parameters and biogeochemical factors on P retention, while the alternative hypothesis (H_1) proposed that physical parameters, specifically flow, transport, and water depth, exert a significantly greater influence on P retention than biogeochemical factors.

Our multi-regression model analyses, supported by high-frequency field data, strongly support the alternative hypothesis. The results indicate that physical parameters, particularly water depth, hydraulic residence time (HRT), and flow path characteristics, are the dominant drivers of P removal in this study. Among these, water depth and inflow TP concentration (TP_{in}) emerged as the most consistent and robust predictors of outflow TP concentration (TP_{out}) across both STA2C3 and STA34C2A. In contrast, simple linear regression revealed a weak or non-significant relationship between TPout and PLR, a key biogeochemical variable, suggesting that PLR alone has limited predictive power.

Additionally, lag time and HRT, both physical transport metrics, effectively accounted for phosphorus residence time, flow delays, and short circuiting within the systems, contributing to improved model performance. The type of vegetation and associated flow structures, such as EAV versus SAV, also played a significant role in shaping the hydrodynamics and, thus, phosphorus retention. Notably, STA34C2A exhibited superior TP removal performance despite its shorter flow path, likely due to the differences in vegetation and hydraulic structure. Furthermore, the field-based estimates of HRT closely matched modeled lag times, providing additional validation for the critical role of physical transport processes in phosphorus retention within constructed wetlands.

4.3. Hypothesis Evaluation

The collective findings of this study refute the null hypothesis (H_0) and provide strong support for the alternative hypothesis (H_1), demonstrating that physical parameters exert a greater influence than biogeochemical factors on P retention in large, constructed wetlands. This dominance of physical drivers was consistent across multiple sites, varying sampling frequencies, and diverse model configurations. Although biogeochemical factors such as the PLR contribute to the overall nutrient dynamics within the system, they were not the primary determinants of P removal efficiency. Instead, physical processes related to water movement, system geometry, and flow resistance, mediated by vegetation type and structural design, were found to be more critical in governing P retention.

These findings have important implications for the design, modeling, and management of large-scale treatment wetlands. First, hydraulic design and vegetation configuration should be prioritized to enhance plug-flow conditions and reduce short circuiting, which can compromise treatment performance. Second, monitoring strategies should incorporate high-resolution hydrologic data to effectively capture temporal variability in flow and transport processes. Finally, the use of lag-based metrics and hydraulic residence time (HRT) estimates derived from data association techniques offers a valuable approach for system assessment, performance optimization, and adaptive management.

5. Conclusions

Our study tested the following hypotheses:

- Null Hypothesis (H_0): There is no significant difference between the influence of physical parameters and biogeochemical factors on phosphorus retention in wetlands.
- Alternative Hypothesis (H_1): Physical parameters (flow, transport, and water depth) have a significantly greater influence on phosphorus retention in wetlands than biogeochemical factors.

Our results provide strong support for the alternative hypothesis (H_1). The analysis clearly demonstrated that physical parameters, particularly HLR, water depth, and HRT, exert a greater and more consistent influence on P removal performance in constructed wetlands than biogeochemical factors like PLR or internal biogeochemical cycling.

Constructed treatment wetlands, especially large STAs, are governed by a mix of controllable (e.g., HLR, water depth, and vegetation type) and uncontrollable (e.g., wildlife activity, storm events, and wind) factors. However, among these, physical and hydraulic controls consistently emerged as the most influential variables shaping TP outflow concentrations across diverse conditions and configurations.

Wetlands characterized by short travel distances and dominated by EAV provided optimum conditions that promote plug-flow hydraulics, and the optimization of HLR and HRT was found to be the most effective in achieving lower TP outflow concentrations. Conversely, in large, single-cell systems dominated by SAV, open water or sparse vegetation, water depth control, and reduction in short circuiting were the most critical physical variables. Despite similar HRT values between the two experiment sites, the wetlands with lower hydraulic conductivity (K) and dense EAV achieved superior P retention, highlighting the importance of hydraulic factors and flow path uniformity.

These findings refute the null hypothesis, as biogeochemical variables such as PLR did not significantly predict TP removal when evaluated alone, and their influence was secondary to hydraulic factors. Even when vegetation type indirectly influenced biogeochemistry, it did so through its primary effect on flow structure and retention time.

Moreover, our study introduced a novel application of data association techniques (cross-correlation analysis) to estimate effective HRT (HRT_P) in large wetland systems, demonstrating strong agreement with dye tracer experiments. This approach offers a practical, data-driven alternative for estimating residence time, which is a key variable in P removal modeling.

We also evaluated the impact of monitoring frequency. While hourly TP sampling revealed variability in outflow concentrations not captured by daily averages, this increased frequency did not improve overall P performance outcomes. However, it emphasized the importance of sample timing in system assessments, suggesting that synchronizing sampling with flow dynamics may be more cost effective than increasing frequency alone.

In summary, the evidence gathered supports the conclusion that physical drivers dominate P retention performance in constructed wetlands. Design and management strategies should prioritize hydraulic control (e.g., HLR, HRT, and water depth) and vegetation structure (EAV vs. SAV) to promote plug-flow conditions and reduce short circuiting. Managing controllable physical parameters offers the greatest potential for optimizing P removal in the face of variable biogeochemical processes and natural disturbances.

Supplementary Materials: The following supporting information can be downloaded at: https://www.mdpi.com/article/10.3390/w17121746/s1, Video S1: Very Fast Water (short circuiting) Field Test 2 092013 V01_Option_1_Video.

Author Contributions: Contributed to conceptualization, M.Z.M. and W.A.M.L.; methodology, M.Z.M. and W.A.M.L.; software, M.Z.M. and W.A.M.L.; validation, M.Z.M. and W.A.M.L.; formal analysis, M.Z.M. and W.A.M.L.; investigation, M.Z.M. and W.A.M.L.; resources, M.Z.M. and W.A.M.L.; data curation, M.Z.M. and W.A.M.L.; writing—original draft preparation, M.Z.M. and W.A.M.L.; writing—review and editing, M.Z.M. and W.A.M.L.; visualization, supervision, M.Z.M. and W.A.M.L.; project administration, M.Z.M. and W.A.M.L. All authors have read and agreed to the published version of the manuscript.

Funding: This research received no external funding.

Data Availability Statement: The raw data supporting the conclusions of this article will be made available by the authors on request.

Acknowledgments: The test would not have been possible without the help from a number of individuals and organizational groups within the SFWMD. The authors would like to acknowledge the support provided by Akintunde Owosina, Bureau Chief of the H&H Bureau, for moving forward with the proposal for the project and making sure it became a reality. We want to thank everyone within the HESM group, the Operations and Maintenance group, and the water managers, control room technicians, and field staff for making it possible to carry out the test over 24-h shifts. We would also like to thank Jeff Kivett of the Operations and Construction Division, Jennifer Leeds of the Policy and Coordination Section, and the field station staff at G-370 and G-372 for prompt responses during the hydraulic field tests.

Conflicts of Interest: The authors declare no conflicts of interest.

References

1. Kadlec, R.H.; Wallace, S.D. *Treatment Wetlands*, 2nd ed.; Taylor & Francis: Boca Raton, FL, USA, 2009.
2. Mitsch, W.J.; Gosselink, J.G. *Wetlands*, 5th ed.; John Wiley & Sons: Hoboken, NJ, USA, 2015.
3. Reddy, K.R.; Kadlec, R.H.; Flaig, E.; Gale, P.M. Phosphorus Retention in Streams and Wetlands: A Review. *Crit. Rev. Environ. Sci. Technol.* **1999**, *29*, 83–146. [CrossRef]
4. Richardson, C.J.; Craft, C.B. Effective phosphorus retention in wetlands: Fact or fiction? In *Constructed Wetlands for Water Quality Improvement*, 1st ed.; CRC Press: Boca Raton, FL, USA, 1993. [CrossRef]
5. Vymazal, J. Removal of nutrients in various types of constructed wetlands. *Sci. Total Environ.* **2007**, *380*, 48–65. [CrossRef]
6. Lal, W.A.M.; Moustafa, M.Z.; Wilcox, W. Field estimation of in-situ vegetation resistance arge managed wetlands using generated waves. In Proceedings of the Environmental & Water Resources Institute (EWRI) of the American Society of Civil, Portland, OR, USA, 1–5 June 2014. [CrossRef]
7. Lal, W.A.M.; Moustafa, M.; Wilcox, W. The use of discharge perturbations to understand in situ vegetation resistance in wetlands: In situ vegetation resistance in wetlands. *Water Resour. Res.* **2015**, *51*, 2477–2497. [CrossRef]
8. Lal, W.A.; Moustafa, M.Z.; Conboy, T.V. Achieving sustainability while maintaining controllability in treatment wetlands. In Proceedings of the AGU, Frontiers in Hydrology Conference, San Juan, PR, USA, 19–24 June 2022.
9. Bachand, P.A.M.; Horne, A.J. Factors affecting nitrogen retention in small constructed wetlands treating agricultural non-point source pollution. *Ecol. Eng.* **2002**, *18*, 351–370. [CrossRef]
10. Bachand, P.A.M.; Horne, A.J. Denitrification in constructed free-water surface wetlands: II. Effects of vegetation and temperature. *Ecol. Eng.* **2000**, *14*, 17–32. [CrossRef]
11. McDowell, R.W.; Haygarth, P.M. Reducing phosphorus losses from agricultural land to surface water. *Curr. Opin. Biotechnol.* **2024**, *89*, 103181. [CrossRef]
12. Moustafa, M.Z.; Lal, W.A.M. Phosphorus Retention in Treatment Wetlands? A Field Experiment Approach: Part 1, Hydrology. *Water* **2025**, *17*, 266. [CrossRef]
13. Arias, C.A.; Del Bubba, M.; Brix, H. Phosphorus removal by sands for use as media in subsurface flow constructed reed beds. *Wat. Res.* **2001**, *35*, 1159–1168. [CrossRef]
14. Arias, C.A.; Brix, H.; Johansen, N.H. Phosphorus removal from municipal wastewater in an experimental two-stage vertical flow constructed wetland system equipped with a calcite filter. *Water Sci. Technol.* **2003**, *48*, 51–58. [CrossRef]
15. Kynkäänniemi, P.; Puustinen, M.; Koskiaho, J.; Tattari, S. Phosphorus retention and release in constructed wetlands: A systematic review. *Environ. Sci. Pollut. Res.* **2021**, *28*, 1641–1657.
16. Rawlik, P. *Evaluation of Sampling Methodologies. Results from Project REST. Remote Environmental Sampling Test*; WR-2017-002; Technical Publication, South Florida Water Management District: West Palm Beach, FL, USA, 2017.

17. Verónica, M.; Eissler, Y.; Fernandez, C.; Cornejo-D'Ottone, M.; Dorador, C.; Bebout, B.M.; Jeffrey, W.H.; Romero, C.; Hengst, M. Greenhouse gases and biogeochemical diel fluctuations in a high-altitude wetland. *Sci. Total Environ.* **2020**, *768*, 144370. [CrossRef]
18. Geranmayeh, P.; Johannesson, K.M.; Ulén, B.; Tonderski, K.S. Particle deposition, resuspension and phosphorus accumulation in small constructed wetlands. *Ambio* **2018**, *47* (Suppl. S1), 134–145. [CrossRef]
19. Reinhardt, M.; Gächter, R.; Wehrli, B.; Müller, B. Phosphorus Retention in Small Constructed Wetlands Treating Agricultural Drainage Water. *J. Environ. Qual.* **2005**, *34*, 1251–1259. [CrossRef]
20. Ury, E.A.; Arrumugam, P.; Herbert, E.R.; Badiou, P.; Page, B.; Basu, N.B. Source or sink? Meta-analysis reveals diverging controls of phosphorus retention and release in restored and constructed wetlands. *Environ. Res. Lett.* **2023**, *18*, 8. [CrossRef]
21. Song, X.; Ehde, P.M.; Weisner, S.E.B. Effects of Water Depth and Phosphorus Availability on Nitrogen Removal in Agricultural Wetlands. *Water* **2019**, *11*, 2626. [CrossRef]
22. Dierberg, F.E.; DeBusk, T.A.; Henry, J.L.; Jackson, S.D.; Galloway, S.; Gabriel, M.C. Temporal and spatial patterns of internal phosphorus recycling in a South Florida (USA) stormwater treatment area. *J. Environ Qual.* **2012**, *41*, 1661–1673. [CrossRef] [PubMed]
23. Kadlec, R.H. Detention and mixing in free water wetlands. *Ecol. Eng.* **1994**, *3*, 345–380. [CrossRef]
24. Moustafa, M.Z. Nutrient retention dynamics of the Everglades Nutrient Removal Project. *Wetlands* **1999**, *19*, 689–704. [CrossRef]
25. Conceição, M.A.; Albuquerque, L.A.; Nogueira, R. Effectiveness and Temporal Variation of a Full-Scale Horizontal Constructed Wetland in Reducing Nitrogen and Phosphorus from Domestic Wastewater. *ChemEngineering* **2018**, *2*, 3. [CrossRef]
26. Baldovi, A.A.; de Barros Aguiar, A.R.; Benassi, R.F.; Vymazal, J.; de Jesus, T.A. Phosphorus removal in a pilot scale free water surface constructed wetland: Hydraulic retention time, seasonality, and standing stock evaluation. *Chemosphere* **2021**, *266*, 128939. [CrossRef]
27. Moustafa, M.Z. Do Wetlands behave Like Shallow Lakes in Terms of Phosphorus Dynamics? *J. Am. Water Resour. Assoc.* **2000**, *36*, 1–12. [CrossRef]
28. Kadlec, R.H. Large Constructed Wetlands for Phosphorus Control: A Review. *Water* **2016**, *8*, 243. [CrossRef]
29. Conn, R.M.; Fiedler, F.R. Increasing Hydraulic Residence Time in Constructed Stormwater Treatment Wetlands with Designed Bottom Topography. *Water Environ. Res.* **2006**, *78*, 2514–2523. [CrossRef] [PubMed]
30. Dierberg, F.E.; DeBusk, T.A. An evaluation of two tracers in surface-flow wetlands: Rhodamine-WT and lithium. *Wetlands* **2005**, *25*, 8–25. [CrossRef]
31. Martinez, C.J.; Wise, W.R. Analysis of constructed treatment wetland hydraulic with the transient storage model OTIS. *Ecol. Eng.* **2003**, *20*, 211–222. [CrossRef]
32. Buchberger, S.G.; Shaw, G.B. An approach toward rational design of constructed wetlands for wastewater treatment. *Ecol. Eng.* **1995**, *4*, 249–275. [CrossRef]
33. Moustafa, M.Z.; Wang, W. Assessment of Wind and Vegetation Interactions in Constructed Wetlands. *Water* **2020**, *12*, 1937. [CrossRef]
34. SFWMD. *South Florida Wading Bird Report 2019*; 2019; Volume 38, Issue 1; pp. 134–145. [CrossRef]
35. South Florida Engineering and Consultants (SFEC). *Sustainable Landscape and Treatment in a Stormwater Treatment Area Study: Final Report (C-4600004949-WO01)*; South Florida Engineering and Consulting, LLC for South Florida Water Management District: West Palm Beach, FL, USA, 2024.

water

MDPI

Review

Multi-Scale Drought Resilience in Terrestrial Plants: From Molecular Mechanisms to Ecosystem Sustainability

Weiwei Lu [1,2], Bo Wu [1,2,*], Lili Wang [1] and Ying Gao [1,2]

1 Institute of Ecological Conservation and Restoration, Chinese Academy of Forestry, Beijing 100091, China; luww@caf.ac.cn (W.L.)
2 Key Laboratory of Desert Ecosystem and Global Change, State Administration of Forestry and Grassland, Beijing 100091, China
* Correspondence: wubo@caf.ac.cn; Tel.: +86-010-62824029

Abstract: Global climate change has intensified the frequency, intensity, and spatial heterogeneity of drought events, posing severe threats to the stability of terrestrial ecosystems. Plant drought resilience, which encompasses a plant's capacity for drought resistance, post-stress recovery, and long-term adaptation and transformation to sustain ecosystem functionality, has emerged as a central focus in botanical and ecological research. This review synthesizes the conceptual evolution of plant drought resilience, from early emphasis on resistance and recovery to the current multi-dimensional framework integrating adaptation and transformation, and synthesizes advances in understanding multi-scale drought resilience in terrestrial plants—spanning molecular, physiological, individual, community, and ecosystem levels. Key mechanisms include molecular/physiological adaptations (osmotic adjustment, antioxidant defense, hydraulic regulation, carbon–water reallocation via gene networks and aquaporins), morpho-anatomical traits (root architectural plasticity, leaf structural modifications, and hydraulic vulnerability segmentation), community/ecosystem drivers (biodiversity effects, microbial symbioses, and soil–plant–feedback dynamics). We critically evaluate quantitative metrics and expose critical gaps, including neglect of stress legacy effects, oversimplified spatiotemporal heterogeneity, and limited integration of concurrent stressors. Future research should prioritize multi-scale and multi-dimensional integrated analysis, long-term multi-scenario simulations with field validation, and harnessing plant–microbe interactions to enhance drought resilience, providing a theoretical basis for ecosystem sustainability and agricultural production under climate change.

Keywords: drought resilience; terrestrial plants; adaptive mechanisms; multi-scale analysis; climate change

1. Introduction

Global climate is undergoing nonlinear evolution characterized predominantly by warming, exhibiting complex features including multi-dimensionality, nonlinearity, and threshold-crossing dynamics [1]. Observational data indicate a 1.09 °C rise in global surface temperature between 1880 and 2022, with the 2013–2022 decade witnessing an accelerated warming rate of 0.38 °C/decade [2]. This warming trajectory intensifies the hydrological cycle, triggering marked increases in the frequency, intensity, and spatial heterogeneity of extreme climate events—manifested as a 56% expansion in regional dry-wet oscillation amplitude (1979–2022) and a 3.2-fold rise in compound extreme event probability [3]. Global climate change intensifies drought dynamics. CMIP6 model simulations project

https://doi.org/10.3390/w17172516

that a global temperature rise of 1.5–4.0 °C will increase potential evapotranspiration across most regions. In East Asia (EAS), for instance, potential evapotranspiration rises by 11.28% under 4.0 °C warming [4]. Drought intensity, coverage area, and duration exhibit significant increases under warming scenarios. In China, anthropogenic climate change has amplified drought frequency by approximately 23%, intensified drought severity by 15%, and expanded affected areas by over 10% [5]. This nonlinear response is particularly pronounced in the Asian monsoon region, where severe drought frequency surges by 30% and duration extends by 20–40% under 2.0 °C warming [6]. At the ecohydrological scale, spatiotemporal reconfiguration of precipitation patterns (e.g., monsoon advancement/delay, rainbelt migration) exacerbates water allocation imbalances, with 34% of global terrestrial ecosystems now experiencing persistent water deficits [7]. Drought emerges as a pivotal threat to terrestrial ecosystem stability under climate change. China, a global drought-stress hotspot, confronts multifaceted challenges of water scarcity and ecological security, directly contributing to a 42% interannual coefficient of variation in degradation rates of northern shelterbelts [8]. Late 20th-century intensive human activities drove vegetation degradation in arid zones exceeding 0.8%/a, catalyzing regional desertification. Implementation of ecological water diversion and Grain-for-Green programs elevated the Normalized Difference Vegetation Index (NDVI) by 0.12–0.18 (2000–2020), significantly restoring ecosystem services [9]. However, synergistic effects between warming-driven glacial retreat acceleration and atmospheric aridification are reducing ecological carrying capacity thresholds and water-use efficiency in arid regions [8].

Drought threatens terrestrial ecosystems through a dual-threat mechanism: directly inducing plant hydraulic failure and carbon starvation, while indirectly elevating pest outbreak risks by 2.7–4.5-fold via altered plant secondary metabolite profiles that compromise chemical defenses [10,11]. Plant drought resilience has emerged as a central focus in botanical and ecological research under climate change, encompassing a plant's capacity for drought resistance, post-stress recovery, and long-term adaptation and transformation to sustain ecosystem functionality and productivity [12,13]. Under global change, vegetation drought resilience emerges as the pivotal regulator of ecosystem sustainability, necessitating elucidation of multi-scale adaptive mechanisms. Current research predominantly focuses on acute drought stress effects (e.g., photosynthetic inhibition, hydraulic dysfunction) and vegetation decline modeled through linear response frameworks [14,15]. Critical knowledge gaps persist; however, including neglect of post-stress recovery dynamics, oversimplification of spatiotemporal heterogeneity in drought–vegetation interactions, and absence of long-term adaptive capacity quantification [16]. These limitations impede holistic understanding of ecological resilience's multidimensional hierarchy, hindering comprehensive assessment of vegetation responses to episodic droughts versus secular climate change.

This review systematically synthesizes recent advances in terrestrial plant drought resilience research to establish a unified conceptual framework spanning molecular, physiological, individual, community, and ecosystem scales. We clarify the conceptual evolution from traditional resistance–recovery paradigms toward multidimensional frameworks integrating adaptation and transformation capacities; comprehensively elucidate key response mechanisms across scales—encompassing molecular, physiological/morphological, and community/ecosystem processes; critically evaluate strengths and limitations of existing quantitative metrics while identifying critical knowledge gaps in spatiotemporal heterogeneity, stress legacy effects, and multi-factor interactions; and propose priority research directions including multi-scale integrative analyses, long-term multi-scenario simulations with field validation, and resilience-enhancing strategies leveraging plant–

microbe interactions. This integrated perspective aims to advance theoretical foundations for climate-resilient ecosystem management and sustainable agriculture.

2. Methodology

This review employed a purpose-driven narrative synthesis to trace the evolution, current trends, and future directions of drought resilience research. Literature was primarily sourced from major academic databases (Web of Science, Scopus, Google Scholar) using core search terms ("drought resilience") and related keywords (e.g., "drought adaptation," "drought resistance"), combined with scale-specific terms (e.g., " micro-scale," "individual scale," "ecosystem," "community," "global"). We prioritized peer-reviewed journal articles (original research, reviews) but included influential meta-analyses, books, and reports from organizations (e.g., IPCC (Intergovernmental Panel on Climate Change)) where critical. Selection emphasized: (a) Recency: Focus on seminal works (pre-2010) and high-impact studies from the last decade (2014–2024), particularly within the last 5 years; (b) Scholarly impact: Citation metrics identified field-defining and widely recognized literature; (c) Spatial scale representation: Studies were systematically categorized into scales to contrast research themes, methods, and interventions; (d) Source rigor: Preference for primary research and systematic reviews. Analyses addressed historical concept/method evolution, scale-specific challenges/strategies, key knowledge gaps, and emerging research frontiers. As a narrative review, this approach allows integrative, cross-scale insights but acknowledges limitations in exhaustive coverage versus systematic methods.

3. Advancing the Conceptualization of Plant Drought Resilience

3.1. Conceptualization and Dimensions of Drought Resilience

The concept of resilience has a longstanding history in ecology, originally describing an ecosystem's capacity to return to its original state following disturbances (Table 1). With intensifying global climate change driving increased frequency and severity of drought events, researchers have progressively incorporated resilience theory into studies of plant drought responses [17]. Initial research primarily centered on resistance (a plant's ability to maintain functionality during drought) and recovery (its capacity to restore functionality after drought cessation) [12,18,19]. Holling [20] established ecological resilience as an ecosystem's capacity to maintain functional and structural stability post-disturbance, introducing two fundamental dimensions: absorptive capacity (threshold of disturbance absorption without structural change, e.g., photosynthetic tolerance under drought) and reorganizational capacity (self-organization into new stable states beyond tipping points, e.g., post-desertification succession). This conceptual foundation, contrasting with engineering resilience (recovery speed), revealed the nonlinear dynamics underpinning drought responses. From the 1990s, Holling's framework evolved into discipline-specific constructs: crop drought resilience and ecological drought resilience [21]. Tilman subsequently decoupled resilience into quantifiable metrics via grassland experiments [22,23]: resistance (functionality maintenance during stress) and recovery (post-stress restoration rate), demonstrating their critical trade-off—a principle guiding drought-tolerant crop breeding. A widely adopted resilience framework conceptualizes resistance and recovery as complementary components of ecosystem adaptability. This concept with its calculation of three resilience indices was proposed in 2011 by Lloret et al. [24]. However, an exclusive focus on resistance and recovery proves insufficient for comprehensively characterizing plants' complex responses to prolonged and recurrent droughts. Schwarz et al. [19] critically evaluated conventional resilience metrics, emphasizing the necessity for more robust and standardized methodologies to assess tree drought responses (Figure 1). Through analyses of divergent drought-response scenarios—including delayed drought responses, pre-drought perturbations, and protracted recovery phases—they demonstrated significant variations in how different metrics capture authentic plant physiological behaviors. Subsequent studies have identified limitations in the Lloret indices, which may yield biased results or misinterpretations under specific conditions. Schwarz et al. [19] examining three common detrended growth scenarios in dendrochronology (Figure 1) revealed deviations from the theoretical assumptions underlying Lloret's original resilience framework. Case analyses demonstrate: (i) drought-induced growth minima occurring after climatic events due to delayed responses (Figure 1a); (ii) violated pre-drought growth assumptions caused by non-drought-related suppression in reference periods (Figure 1b); and (iii) protracted recovery from sequential droughts necessitating both multi-year drought data aggregation and post-drought reference period re-calibration (Figure 1c). Subsequently, drought resilience has emerged as a comprehensive conceptual framework that extends beyond resistance and recovery to encompass a plant's capacity for adaptive adjustment and structural transformation in response to persistent drought stress, thereby ensuring sustained ecological functionality [12,25]. This conceptual evolution reflects the research community's deepening understanding of the intricate dynamics in plant–environment interactions. Post-2000, accelerated by climate change, drought resilience has been systematically defined with progressively enriched dimensions. In this section, where applicable, authors are required to disclose details of how generative artificial intelligence (GenAI) has been used in this paper (e.g., to generate text, data, or graphics, or to assist in study design, data collection, analysis, or interpretation). The use of GenAI for superficial text editing (e.g., grammar, spelling, punctuation, and formatting) does not need to be declared.

Table 1. Historical evolution of key conceptual and methodological advances in plant drought resilience research.

Stage	Timeframe	Key Advances	Significance
Conceptual Foundation	1970s~90s	Focus on drought resistance traits: osmoregulation, deep rooting, stomatal control; Distinction between avoidance vs. tolerance.	Established physiological basis of drought survival.
Paradigm Shift to Resilience	Early 2000s~2010s	Resilience triad: Resistance + Recovery + Elasticity; Hydraulic failure-carbon starvation model; Drought memory concept.	Shifted focus to post-stress recovery dynamics; linked physiology to mortality.
Multi-Omics Integration	2010s–Present	Transcriptomic networks; CRISPR (Clustered Regularly Interspaced Short Palindromic Repeats)-based resilience enhancement; PGPR (Plant Growth-Promoting Rhizobacteria)-mediated rhizosphere resilience; UAV (Unmanned Aerial Vehicle) phenomics.	Decoded molecular to ecosystem-scale mechanisms; enabled high-throughput screening.
Application and Scaling		Resilience-prioritized breeding; Microbial bioinoculants; AI-driven irrigation models; Mixed-species restoration.	Bridged theory to field practice; optimized resource use.

3.2. Drought Resilience Indicators

Drought resilience indicators serve as critical tools for assessing the ability of ecosystems, agricultural systems, and even socio-economic systems to maintain functionality and recover from drought disturbances. These indicators span multiple dimensions—from plant physiological levels to regional scales—and are designed to quantify a system's resistance to drought, recovery capacity, and overall adaptation capabilities [26,27]. An in-depth understanding of these indicators is crucial for developing effective drought response strategies and ensuring regional sustainable development [2,28]. Resistance refers to a system's ability to withstand negative impacts and preserve structure/function during drought [26,29]. For instance, highly resistant systems exhibit minimal initial impact, such as drought-tolerant maize varieties maintaining high germination vigor, germination rates, and growth rates of coleoptiles and roots under water stress during seedling stages [30]. Quantitatively, plant resistance can be measured by reduced growth or productivity decline during drought, often expressed as the ratio of growth under drought to growth under favorable conditions [31]. Recovery denotes the ability and speed of a system to return to its pre-drought state after drought cessation [25,26,29]. This is typically assessed by monitoring post-drought recovery trajectories of ecosystem functions (e.g., vegetation productivity via satellite remote sensing), where prolonged recovery may predict increased future mortality risks [25,29,32].

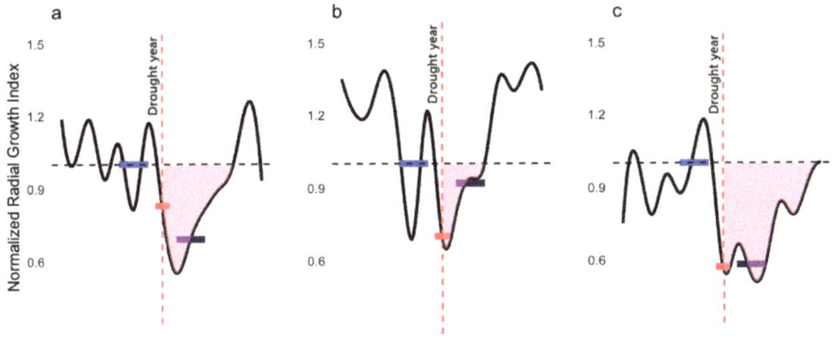

Lloret & Proposed Indices (a)

Indices	Lloret	Proposed
Resilience	0.69	
Recovery	0.83	
Resistance	0.83	
Total growth reduction		1.07
Recovery period		5
Average growth reduction		0.27
Average recovery rate		20 %

Lloret & Proposed Indices (b)

Indices	Lloret	Proposed
Resilience	0.92	
Recovery	1.32	
Resistance	0.7	
Total growth reduction		0.53
Recovery period		3
Average growth reduction		0.18
Average recovery rate		33.3 %

Lloret & Proposed Indices (c)

Indices	Lloret	Proposed
Resilience	0.58	
Recovery	1.02	
Resistance	0.57	
Total growth reduction		2.10
Recovery period		7
Average growth reduction		0.30
Average recovery rate		13.9 %

Figure 1. Reillustration of scenarios where Lloret's resilience indices may yield incomplete or misleading assessments of drought response (modified from [19]). Subfigures depict: (**a**) Lagged growth minimum occurring in year t + 1 following drought at year; (**b**) Significant pre-drought growth suppression within the reference period; (**c**) Multi-year drought events leading to prolonged recovery. Calculations (based on simulated detrended data with pre-drought growth set to 1) consider a 3-year pre-drought (blue lines) and post-drought (black lines) reference period. Red lines indicate growth in the drought year.

Beyond resistance and recovery, drought resilience includes adaptation—a system's capacity to adjust its structure or function for improved future drought response—and transformation, which involves fundamental system changes to address long-term or extreme climate shifts, moving beyond mere recovery toward more sustainable states [19,33] (Figure 2). Figure 2 conceptualizes drought resilience dimensions through comparative vegetation dynamics. The hierarchy progresses from overarching Adaptation (evolutionary/physiological adjustments) to core Drought Resilience (integrated stress response capacity), which bifurcates into Resistance (stress-withstanding ability via osmotic regulation, hydraulic control) and Recovery (post-stress function restoration via metabolic reprogramming). Trees/Ecosystems A and B exemplify divergent resilience strategies: A demonstrates high resistance but limited recovery (e.g., deep-rooted species maintaining function during drought yet slow to rebound), while B exhibits lower resistance but rapid recovery (e.g., drought-deciduous species sacrificing immediate function for efficient post-rain regrowth). The bidirectional arrow signifies that resistance–recovery tradeoffs in real-world systems are dynamically mediated by species traits, soil properties, and historical climate exposure. Though adaptation and transformation metrics are less frequently quantified in empirical studies compared to resistance and recovery, they remain essential for understanding the macro-level evolution of drought resilience [34,35].

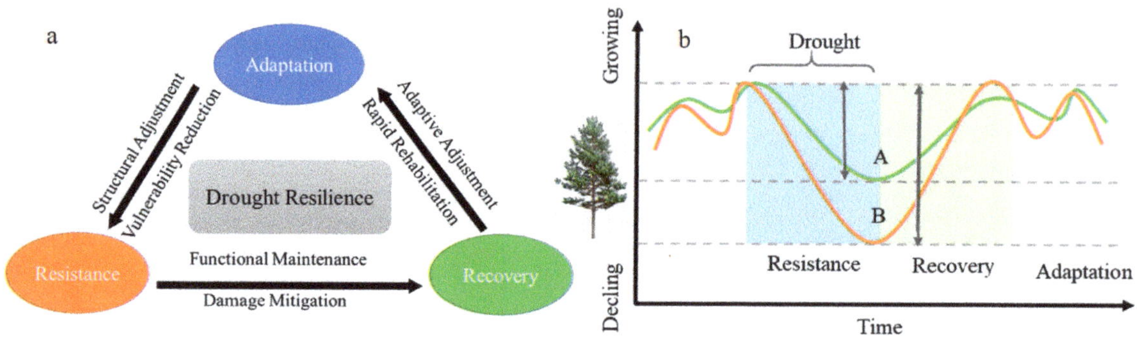

Figure 2. Dimensions of Drought Resilience. Schematic of vegetation growth dynamics and drought resilience responses under drought events. (**a**) Conceptual framework of key resilience strategies; (**b**) Temporal dynamics of resistance and recovery in two distinct systems (A and B). A and B refer to two trees or ecosystems.

4. Plant Drought Resilience Across Spatial Scales

4.1. Molecular and Cellular Mechanisms

4.1.1. Water Relations Regulation

Under drought stress, plants enhance drought resistance through multi-level synergistic mechanisms. Plants optimize water utilization via stomatal behavior and osmotic adjustment: Isohydric species strictly limit stomatal aperture to avoid hydraulic failure, while anisohydric species permit tissue water potential decline to sustain carbon assimilation [36]. Concurrently, plants accumulate osmolytes (e.g., proline, betaine) to lower cellular osmotic potential, ensuring turgor maintenance [37]. Molecular studies reveal that abscisic acid (ABA) signaling mediates rapid stomatal closure by regulating ion channels in guard cells, with transcription factors (e.g., DREB (Dehydration-Responsive Element-Binding protein)) fine-tuning stomatal dynamics for water balance [38].

Plants employ osmotic adjustment to maintain cellular turgor and water homeostasis during drought, a critical physiological adaptation achieved through three synergistic mechanisms: (1) Preservation of plasma and endosomal membrane integrity to prevent turgor loss [39]; (2) Accumulation of compatible solutes that reduce cellular osmotic potential, thereby enhancing soil water extraction efficiency [40]; and (3) Responsiveness to exogenous interventions exemplified by nitric oxide (NO) application, which elevates shoot length (66.60%), root length (29.38%), shoot dry weight (26.69%), and root dry weight (28.52%), while concurrently boosting photosynthesis, antioxidant defenses, and osmotic adjustment to confer comprehensive drought resilience [9].

4.1.2. Carbon–Water Coupling Balance

Plants maintain carbon–water equilibrium through photosynthesis-transpiration decoupling and carbon allocation: C4 (C4 carbon fixation pathway) and CAM (Crassulacean Acid Metabolism) plants enhance water use efficiency (WUE) by spatially or temporally separating CO_2 fixation from transpiration. Despite drought-induced photosynthesis suppression leading to carbon starvation, plants prioritize carbon allocation to critical organs (e.g., roots) and leverage non-structural carbohydrate (NSC) reserves as metabolic substrates and osmoregulators [41], sustaining survival under limited carbon supply. Advances in molecular biology and genomics have propelled in-depth research into drought resilience-related genes. Studies reveal that specific genes—such as transcription factors like DREB and NAC—play critical roles in regulating plant drought responses by controlling water management and carbon metabolism pathways. For instance, gene-editing technologies like CRISPR are being deployed to enhance plant drought tolerance [42].

4.1.3. Oxidative Stress Defense

The plant antioxidant defense system primarily protects cells from oxidative damage by balancing the clearance of reactive oxygen species (ROS), involving multiple biochemical processes. This mechanism is particularly crucial under environmental stresses such as drought or light stress. To counter drought-triggered reactive oxygen species (ROS) accumulation, plants activate multi-tiered protection: Antioxidant enzymes (e.g., SOD (Superoxide Dismutase), CAT (Catalase)) scavenge ROS to safeguard membrane integrity [43], while heat shock proteins (HSPs) stabilize denatured proteins to maintain cellular homeostasis [44].

4.2. Leaf-Level Responses

4.2.1. Physiological and Biochemical Responses

Photosynthesis, the cornerstone of plant growth and productivity, exhibits drought-driven dynamics that directly determine resistance capacity and post-stress recovery resilience. Under severe or prolonged drought—and co-occurring waterlogging—photosynthetic impairment manifests through reduced stomatal conductance, disrupted electron transport, and Rubisco inactivation [45,46]. While most photosynthetic parameters recover after rehydration, the extent and velocity of recovery demonstrate significant genotypic variation and scale inversely with stress intensity [45]. Critically, persistent photochemical damage serves as a threshold metric for resilience, where sustained depression of the photosystem II (PSII) maximum quantum yield (F_v/F_m) beyond 15% post-rehydration predicts elevated mortality risk during subsequent droughts in tropical rainforest species [38]. This legacy effect confirms that drought impacts transcend transient physiological disruption, driving cumulative functional deterioration that compromises long-term plant performance.

4.2.2. Photosynthetic Regulation

Physiologically, drought reduces leaf relative water content (RWC)—a critical water status indicator [47]. Stomatal regulation minimizes water loss through closure but concurrently limits photosynthesis and growth [48–50]. Under drought conditions, plants adjust carbon metabolism and allocation strategies. Research indicates that while drought suppresses photosynthesis and induces carbon starvation, plants counteract this challenge by preferentially allocating carbon resources to critical organs such as roots [38]. Through reducing carbon investment in non-essential tissues and organs, plants sustain vital functions under constrained carbon supply. Osmotic adjustment counters low water potential via proline, soluble sugars, and mannitol accumulation, preserving cell turgor and function [47]. To mitigate drought-induced reactive oxygen species (ROS), plants elevate antioxidant enzymes (SOD, CAT, POD (Peroxidase)) and non-enzymatic antioxidants (glutathione, vitamin C) [51–53]. Photosynthesis declines markedly, evidenced by reduced net photosynthetic rate, transpiration, and stomatal conductance [48,54]. Regulation of carbon metabolism and carbon reserves plays a critical role in plant drought resilience, particularly under prolonged drought stress. Under drought conditions, plants adjust carbon metabolism and allocation strategies. Research indicates that while drought suppresses photosynthesis and induces carbon starvation, plants counteract this challenge by preferentially allocating carbon resources to critical organs such as roots [38]. Through reducing carbon investment in non-essential tissues and organs, plants sustain vital functions under constrained carbon supply.

4.3. Whole-Plant and Species-Level Traits

4.3.1. Morphological Adaptations

Under drought stress, plants undergo significant morphological changes to enhance drought resilience. Root architecture—the key organ for water uptake—adapts through increased root-to-shoot ratio and development of deeper/extensive root systems, improving water absorption efficiency [55,56]. Plant drought resilience is fundamentally governed by interconnected root architectural traits and rhizosphere processes, with root system functionality serving as the primary determinant of water acquisition efficiency under moisture deficit [57]. Adaptive plasticity in root morphology and physiology during drought-rehydration cycles critically drives climate resilience and productivity optimization, though such plasticity can be compromised by concurrent stresses—exemplified by cadmium contamination disrupting root circadian rhythms and suppressing lateral root development [39,57]. Notably, steep lateral root angles (LRA) enabling deeper soil exploration significantly enhance drought resilience through access to subsoil water reserves, as demonstrated by the superior drought survival of narrow-LRA cotton genotypes [58]. Furthermore, drought-adaptive rhizosheath formation—mediated by cultivar-specific biophysical and biochemical traits—orchestrates plant–soil–microbe interactions, where root exudate-driven communications constitute underexplored pathways for drought mitigation [59,60]. Aboveground, plants reduce height, decrease leaf area, or shed leaves to minimize transpirational water loss [49,61], exemplified by eucalyptus seedlings adjusting photosynthesis and growth post-water deficit pretreatment [62].

4.3.2. Microbial and Metabolic Contributions

Plant–microbe interactions and endogenous metabolic processes constitute critical determinants of drought resilience. Rhizospheric microbiomes manifest drought-responsive adaptations, as evidenced by wheat rhizospheres where stress-enriched microbial consortia and metabolites enhance drought tolerance through plant–microbe crosstalk, exhibiting persistent legacy effects [63]. Arbuscular mycorrhizal fungi (AMF) significantly potentiate plant drought resilience via symbiotic water-nutrient exchange [64], while plant growth-promoting rhizobacteria (PGPR) synergistically reinforce resistance to combined drought–salinity stressors [65]. Metabolically, trehalose-6-phosphate (T6P) signaling precursor treatment demonstrates efficacy in boosting spring wheat yield and drought resilience through osmotic regulation [66].

4.3.3. Hydraulic Traits

Hydraulic traits critically govern drought survival. Xylem embolism—gas bubble formation in conduits—impedes water transport and can cause mortality. Plants combat this through anatomical adaptations like modified vessel diameter/density to enhance hydraulic safety [67]. A key strategy is hydraulic vulnerability segmentation, where distal organs (e.g., leaves, fine roots) are sacrificed before core structures, preserving overall hydraulic integrity [68]. Anatomical modifications further optimize drought resistance. Thickened leaf cuticles reduce non-stomatal water loss and improve water-use efficiency [69]. Vascular bundle organization—including xylem vessel size, density, and arrangement—determines hydraulic efficiency and embolism resistance. Drought-tolerant species often exhibit smaller vessel diameters and higher vessel density to enhance hydraulic safety [70].

4.4. Community and Ecosystem Dynamics

4.4.1. Plant Biodiversity and Community Properties

At the ecosystem scale, biodiversity exerts critical influences on drought resilience. Hydraulic diversity—particularly the variability in plant functional traits related to water transport—serves as a key modulator of land–atmosphere feedbacks during drought events, thereby regulating ecosystem resilience [71]. Community composition and species diversity determine ecosystem resistance and recovery capacity, with warming and nitrogen deposition demonstrating drought-contingent effects on grassland resilience through compositional shifts [72,73]. Precipitation anomalies (including drought and pluvial episodes) further reconfigure productivity resilience via altered community dynamics, evidenced by drought-avoidant seed dormancy mechanisms in annual plant communities where wet seasons counteract prior drought impacts [74].

4.4.2. Trait-Based Adaptive Strategies in Community Drought Response

Drought resilience manifests through changes in community composition, functional traits, and species interactions [75]. Ecological theory predicts that community assembly processes may lead to functional traits of coexisting species being more divergent ("divergence") or more similar ("convergence") than expected by random chance [76]. Backhaus et al. [77] aimed to identify functional trait patterns across spatial scales and successional stages, discussing their underlying community assembly mechanisms. This implies that under varying drought conditions, communities adaptively adjust their trait composition to maintain stability and functionality. For instance, under prolonged drought stress, communities may selectively favor species with enhanced drought-tolerant traits, resulting in trait convergence phenomena.

4.4.3. Biodiversity Dimensions Mediating Ecosystem Resilience

Species diversity, particularly the diversity of functional traits, is recognized as a critical factor enhancing community drought resilience [78–80]. Research indicates that the positive effect of species richness on community drought resistance may be modulated by species evenness, operating through altered complementarity effects [81]. In forest communities, the coexistence of species with divergent drought-resistance strategies can bolster ecosystem-wide stability [13,78]. For instance, studies have shown that increased abundance of drought-tolerant species may compensate for declines in forest productivity by enhancing richness in drought-tolerant traits [82]. However, other research suggests that in multispecies grasslands, drought tolerance is more strongly determined by species identity and functional group diversity than by simple species diversity [83].

4.4.4. Successional Dynamics and Biotic–Abiotic Interactions

Community composition and functional traits undergo changes during succession to adapt to environmental pressures. Under drought conditions, trait convergence may signify that communities enhance overall resilience through selective retention of drought-tolerant species [77]. Interactions among species—including competitive and facilitative effects—can significantly influence community responses to drought [84]. In tropical seedling communities, neighboring diversity modulates seedling drought resistance [85]. Interactions between soil water and plants, along with their legacy effects, also play pivotal roles in drought resistance and resilience within arid ecosystems [33].

5. Quantification and Assessment Methods for Plant Drought Resilience

5.1. Multidimensional Metrics for Plant Drought Resilience Assessment

Current research employs diverse metrics and methodologies to evaluate plant drought resilience from multiple perspectives [19,83,86]. Physiological, biochemical, morphological, and resilience indices are critical for quantifying plant drought resilience (Table 2). Physiological indicators include water status markers, osmoprotectants (proline, soluble sugars), membrane stability (CMS), photosynthetic efficiency (Pn, F_v/F_m), antioxidant enzymes, and stomatal regulation. Morphological traits encompass biomass, root architecture, and leaf structure. Resilience is evaluated via Lloret [24] indices (resistance/recovery/resilience), satellite-derived productivity metrics (NDVI (Normalized Difference Vegetation Index), EVI (Enhanced Vegetation Index), GPP (Gross Primary Productivity)), and integrated statistical methods (PCA (Principal Component Analysis), membership functions). Multi-dimensional assessment enables robust analysis of plant adaptive capacity under drought stress.

Table 2. Classification framework for multi-dimensional assessment metrics of plant drought resilience. Integrating physiological, biochemical, morphological, and comprehensive resilience indicators.

Category	Indicator	Description
Physiological and Biochemical	Relative Water Content (RWC)	Measures plant tissue water status; declines under drought; correlates with drought resistance.
	Osmotic Potential (Ψ_s)	Quantifies osmotic adjustment capacity; critical for maintaining turgor under dehydration.
	Proline (Pro) and Soluble Sugar (SS)	Osmoprotectants maintain cell turgor and macromolecule integrity; accumulate during osmotic adjustment.
	Cell Membrane Stability (CMS)	Reflects membrane damage resistance; higher CMS indicates superior drought tolerance.
	Photosynthetic Metrics (Pn, F_v/F_m, φPSII)	Indicate light-use efficiency and photosynthetic apparatus integrity under drought.
	Antioxidant Enzyme Activity	Scavenges ROS to mitigate oxidative damage; enhanced under drought.
	Stomatal Conductance (gs)	Real-time gas exchange regulator; rapid decline triggers drought avoidance.
	Stomatal Density (SD)	Structural adaptation trait; higher density enhances gas exchange flexibility.
Hydraulic Traits	Water Potential (Ψ)	Leaf/cellular water status indicator; negative values intensify with drought severity.
	Xylem Embolism Resistance (P_{50})	Critical threshold of water potential causing 50% hydraulic conductivity loss; key survival trait.
	Leaf Turgor Loss Point (PTLP)	Osmotic potential at cell turgor loss; determines stomatal closure threshold.
	Specific Hydraulic Conductivity (Ks)	Mass-specific xylem efficiency: conductivity per unit xylem area under pressure gradient.
Morphological	Biomass and Growth (height, leaf area)	Drought suppresses growth, reducing biomass and morphological parameters.
	Root Morphology (e.g., root-to-shoot ratio)	Deep/extensive roots enhance water uptake; root-to-shoot ratio indicates resource allocation.
	Leaf Anatomy (thickness, area)	Influences water-use efficiency and transpiration rates.

Table 2. *Cont.*

Category	Indicator	Description
Resilience Indices	Lloret Indices (Resistance/Recovery/Resilience)	Quantifies system stability during disturbance (resistance), post-stress recovery, and overall functional maintenance.
	Productivity-Based Metrics (NDVI/EVI/GPP)	Satellite-derived indices assessing ecosystem-scale resistance and recovery.
	Integrated Methods (PCA, membership function)	Statistical approaches combining multiple indicators for holistic resilience evaluation.

Tree-ring analysis serves as the primary technique for quantifying resilience patterns through tree-ring width (TRWi) and density indices. In a representative study, Kang et al. [87] analyzed *Betula platyphylla* in semi-arid northern China, defining resistance as growth reduction magnitude during drought and resilience as post-disturbance recovery rate. This approach is constrained by its reliance on manual sampling for high-resolution data and limited capacity for capturing short-term dynamic responses. Notably, competition effects (e.g., neighbor interference) significantly modulate local resilience: increased stand density reduces individual tree resistance while enhancing resilience, necessitating quantification via mixed-effects models (as validated by a 166-case meta-analysis) [88]. Integrated remote sensing and ecological indices enable regional resilience mapping through satellite data (e.g., NDVI, GPP) coupled with ground validation. Zhang et al. [27] combined Leaf Area Index (LAI) with the Standardized Precipitation Evapotranspiration Index (SPEI), revealing that drought typology (meteorological vs. agricultural) governs vegetation response patterns, where drought timing (dry vs. wet seasons) critically alters recovery trajectories. Cross-scale evidence confirms biodiversity enhances resilience: Bai and Tang's [89] analysis of 4072 global sites verified that species richness elevates resistance and resilience through functional redundancy and resource partitioning mechanisms, though climate zone dependencies (humid vs. arid regions) require differential validation.

5.2. Limitations and Emerging Directions in Plant Drought Resilience Assessment

However, these assessment approaches have certain limitations. For instance, physiological indicators are often sensitive to short-term environmental fluctuations, making it difficult to reflect the long-term drought adaptation of plants. Morphological traits, on the other hand, are relatively stable but have a slow response speed, which may lag behind the actual drought stress process. In recent years, with the development of molecular biology techniques, molecular markers have gradually been incorporated into the quantitative evaluation system of plant drought resilience. Genes related to drought resistance, such as those encoding dehydrins and transcription factors involved in stress signaling pathways, their expression levels and polymorphisms can provide new insights into the genetic basis of plant drought resilience [37,65]. Metabolomics analysis, which identifies and quantifies the changes in small molecule metabolites under drought stress, can also complement the traditional physiological and biochemical indicators, revealing the metabolic regulatory networks underlying plant drought adaptation.

Another important aspect is the integration of assessments at different scales. From the cellular level to the whole plant, and further to the community and ecosystem levels, there are complex interactions and feedbacks. For example, the drought resilience of individual plants may affect the structure and function of the entire community, and the microclimate and soil conditions of the community can in turn influence the drought response of individual plants [21,25,45]. Therefore, developing a multi-scale integrated evaluation framework is crucial for a comprehensive understanding of plant drought resilience.

In practical applications, there are also challenges in the quantitative evaluation of plant drought resilience. One of them is the lack of unified standards and criteria for different indicators and methods, which makes it difficult to compare and integrate results from different studies. For example, the threshold values of physiological indicators for different plant species under drought stress may vary greatly, and there is no universal reference standard. In addition, the cost and complexity of some advanced techniques, such as metabolomics and high-throughput sequencing, limit their widespread application in large-scale field studies.

6. Conclusions and Future Perspectives

Plant responses to drought stress represent a critical challenge for their survival and growth, particularly under global climate change where the intensity and frequency of drought events are escalating, posing severe threats to agricultural production and ecosystem stability [35,90,91]. To adapt to drought, plants employ complex multiscale and hierarchical mechanisms spanning morphological, physiological, biochemical, hydraulic, and anatomical traits [48,92,93]. A deeper understanding of these drought resilience mechanisms is essential for developing drought-tolerant crops and advancing ecological restoration. Despite significant advances in understanding plant drought resilience at the individual scale, key limitations persist. Current limitations in plant drought resilience research:

(a) Lack of Standardized Metrics and Scale Effects: Inconsistent definitions and quantification methods for "resilience" across scales hinder cross-study comparisons. Understanding how micro-scale mechanisms (e.g., leaf physiology, gene expression) translate to macro-scale stability remains unclear.
(b) Oversimplified Drought Characterization: Treating drought as a single event neglects the independent and combined impacts of key dimensions like severity, duration, and timing (seasonality). Modeling these complex interactions at large scales is challenging.
(c) Neglect of Long-Term Effects and Recovery Dynamics: Focus remains on short-term resistance and immediate recovery. Understanding cumulative stress effects from repeated droughts, full recovery timelines/mechanisms, and global trends in the resistance–recovery trade-off is insufficient.
(d) Complexity of Soil–Microbiome Interactions: While soil microbes (e.g., AMF) are known to enhance drought resilience, predicting how soil depth, plant community composition, and highly diverse, dynamic microbial communities collectively influence resilience at ecosystem scales is a major unresolved challenge.
(e) Quantifying Diversity's Role: Precisely measuring how different types of plant diversity (species, functional, genetic) contribute to resilience and integrating this into large-scale models is difficult. Simply increasing species number may not suffice; functional complementarity and the ecosystem-level impacts of genetic engineering need deeper investigation.
(f) Data and Modeling Constraints: Scarcity of high-resolution, long-term ecosystem flux data limits fine-scale drought response analysis. Global hydrological resilience assessment needs better data/methods. Models struggle to capture complex drought dimension interactions and lack validation due to limited field control and long-term data.

Future research should target the identified gaps and limitations, with priorities on the following:

(a) Integrated Multi-Scale Methodologies: Advancing frameworks that synergize physiological, biochemical, and molecular regulation through multi-omics approaches to identify resilience biomarkers, while integrating ecohydrology with the Soil–Plant–Atmosphere Continuum (SPAC) to quantify hydraulic traits and stomatal dynamics across drought types. Development of novel metrics for resistance (stress-withstanding capacity) and recovery (post-stress restoration) remains essential.
(b) Long-Term Experimental Validation: Implementing multi-year, multi-scenario drought simulations to assess ecosystem resilience, with emphasis on rainfall pattern variations and Plant–Soil Feedback (PSF) dynamics. Research should elucidate how drought severity and historical precipitation modulate PSF-driven community assembly and ecosystem recovery.
(c) Harnessing Plant–Microbe Synergies: Prioritizing rhizosphere microbiome engineering for staple crops to quantify trait relationships and legacy effects. Synergistic applications of bioengineered microbes, plant growth regulators (PGRs), and targeted miRNA editing warrant mechanistic exploration to enhance drought tolerance.

In summary, future research on plant drought resilience will constitute a comprehensive research framework that integrates multidisciplinary approaches and combines macro- and micro-level perspectives. By deepening our understanding of physiological, biochemical, and molecular mechanisms, leveraging advanced experimental methods and data analysis techniques, and harnessing the potential of plant–microbe interactions, we will be better equipped to address drought challenges posed by climate change, thereby securing global food security and ecosystem health.

Author Contributions: Conceptualization, W.L. and B.W.; methodology, B.W.; software, W.L.; formal analysis, W.L.; investigation, W.L.; data curation, L.W.; writing—original draft preparation, W.L.; writing—review and editing, B.W.; supervision, Y.G.; project administration, B.W.; funding acquisition, W.L. All authors have read and agreed to the published version of the manuscript.

Funding: This research was funded by Chinese Academy of Forestry [CAFYBB2024MA043] and National Key R&D Program of China [2023YFF1305303-01].

Data Availability Statement: The data that support the findings of this study are available on request from the corresponding author upon reasonable request.

Conflicts of Interest: The authors declare no conflicts of interest.

References

1. IPCC. *Climate Change 2021: The Physical Science Basis*; IPCC: Geneva, Switzerland, 2021.
2. Forster, P.M.; Smith, C.; Walsh, T.; Lamb, W.F.; Lamboll, R.; Hall, B.; Hauser, M.; Ribes, A.; Rosen, D.; Gillett, N.P. Indicators of Global Climate Change 2023: Annual update of key indicators of the state of the climate system and human influence. *Earth Syst. Sci. Data* **2024**, *16*, 2625–2658. [CrossRef]
3. Perret, D.L.; Evans, M.E.; Sax, D.F. A species' response to spatial climatic variation does not predict its response to climate change. *Proc. Natl. Acad. Sci. USA* **2024**, *121*, e2304404120. [CrossRef]
4. Ji, Y.; Fu, J.; Lu, Y.; Liu, B. Three-dimensional-based global drought projection under global warming tendency. *Atmos. Res.* **2023**, *291*, 106812. [CrossRef]
5. Zhao, R.; Sun, H.; Xing, L.; Li, R.; Li, M. Effects of anthropogenic climate change on the drought characteristics in China: From frequency, duration, intensity, and affected area. *J. Hydrol.* **2023**, *617*, 129008. [CrossRef]
6. Kim, J.; So, J.; Bae, D. Global warming impacts on severe drought characteristics in Asia monsoon region. *Water* **2020**, *12*, 1360. [CrossRef]
7. Vacek, Z.; Vacek, S.; Cukor, J. European forests under global climate change: Review of tree growth processes, crises and management strategies. *J. Environ. Manag.* **2023**, *332*, 117353. [CrossRef]
8. Zhang, Z.; Zhang, L.; Xu, H.; Creed, I.F.; Blanco, J.A.; Wei, X.; Sun, G.; Asbjornsen, H.; Bishop, K. Forest water-use efficiency: Effects of climate change and management on the coupling of carbon and water processes. *For. Ecol. Manag.* **2023**, *534*, 120853. [CrossRef]

9. Liu, Y.; Li, P.; Wang, Y.; Xu, X. Quantification of carbon-water dynamics in soil-perennial grass (*Bothriochloa ischaemum*) feedbacks under drought stress following a double isotope-labelled pulse experiment. *Agric. For. Meteorol.* **2023**, *329*, 109270. [CrossRef]
10. Jactel, H.; Moreira, X.; Castagneyrol, B. Tree diversity and forest resistance to insect pests: Patterns, mechanisms, and prospects. *Annu. Rev. Entomol.* **2021**, *66*, 277–296. [CrossRef]
11. Hodgson, D.; McDonald, J.L.; Hosken, D.J. What do you mean, 'resilient'? *Trends Ecol. Evol.* **2015**, *30*, 503–506. [CrossRef]
12. Volaire, F. A unified framework of plant adaptive strategies to drought: Crossing scales and disciplines. *Glob. Change Biol.* **2018**, *24*, 2929–2938. [CrossRef] [PubMed]
13. Yao, Y.; Liu, Y.; Song, J.; Tao, S.; Li, Y.; Wu, T.; Wang, Y.; Wang, S.; Fu, B. Declining tradeoff between resistance and resilience of ecosystems to drought. *Earth's Future* **2024**, *12*, e2024EF004665. [CrossRef]
14. Smith, T.; Traxl, D.; Boers, N. Empirical evidence for recent global shifts in vegetation resilience. *Nat. Clim. Change* **2022**, *12*, 477–484. [CrossRef]
15. Fu, B.; Stafford-Smith, M.; Wang, Y.; Wu, B.; Yu, X.; Lv, N.; Ojima, D.S.; Lv, Y.; Fu, C.; Liu, Y. The Global-DEP conceptual framework—Research on dryland ecosystems to promote sustainability. *Curr. Opin. Environ. Sustain.* **2021**, *48*, 17–28. [CrossRef]
16. Kannenberg, S.A.; Driscoll, A.W.; Szejner, P.; Anderegg, W.R.; Ehleringer, J.R. Rapid increases in shrubland and forest intrinsic water-use efficiency during an ongoing megadrought. *Proc. Natl. Acad. Sci. USA* **2021**, *118*, e2118052118. [CrossRef]
17. Yu, H.; Ma, Q.; Liu, X.; Li, Y.; Li, L.; Qi, M.; Wu, W.; Wang, Y.; Xu, Z.; Zhou, G. Resistance, recovery, and resilience of desert steppe to precipitation alterations with nitrogen deposition. *J. Clean. Prod.* **2021**, *317*, 128434. [CrossRef]
18. Miao, Y.; Zhou, Z.; Jiang, M.; Song, H.; Yan, X.; Liu, P.; Ji, M.; Han, S.; Chen, A.; Wang, D. Resistance and resilience of nine plant species to drought in Inner Mongolia temperate grasslands of northern China. *Appl. Sci.* **2022**, *12*, 4967. [CrossRef]
19. Schwarz, J.; Skiadaresis, G.; Kohler, M.; Kunz, J.; Schnabel, F.; Vitali, V.; Bauhus, J. Quantifying growth responses of trees to drought—A critique of commonly used resilience indices and recommendations for future studies. *Curr. For. Rep.* **2020**, *6*, 185–200. [CrossRef]
20. Holling, C.S. *Resilience and Stability of Ecological Systems*; Annual Reviews: San Mateo, CA, USA, 1973.
21. Rodriguez-Iturbe, I. Ecohydrology: A hydrologic perspective of climate-soil-vegetation dynamies. *Water Resour. Res.* **2000**, *36*, 3–9. [CrossRef]
22. Tilman, D.; Downing, J.A. Biodiversity and stability in grasslands. *Nature* **1994**, *367*, 363–365. [CrossRef]
23. Tilman, D.; Reich, P.B.; Knops, J.M. Biodiversity and ecosystem stability in a decade-long grassland experiment. *Nature* **2006**, *441*, 629–632. [CrossRef]
24. Lloret, F.; Keeling, E.G.; Sala, A. Components of tree resilience: Effects of successive low-growth episodes in old ponderosa pine forests. *Oikos* **2011**, *120*, 1909–1920. [CrossRef]
25. Sykes, A.J.; Macleod, M.; Eory, V.; Rees, R.M.; Payen, F.; Myrgiotis, V.; Williams, M.; Sohi, S.; Hillier, J.; Moran, D. Characterising the biophysical, economic and social impacts of soil carbon sequestration as a greenhouse gas removal technology. *Glob. Change Biol.* **2020**, *26*, 1085–1108. [CrossRef] [PubMed]
26. Deepa, R.; Anandhi, A.; Alhashim, R. Volumetric and impact-oriented water footprint of agricultural crops: A review. *Ecol. Indic.* **2021**, *130*, 108093. [CrossRef]
27. Zhang, Y.; Liu, X.; Jiao, W.; Wu, X.; Zeng, X.; Zhao, L.; Wang, L.; Guo, J.; Xing, X.; Hong, Y. Spatial heterogeneity of vegetation resilience changes to different drought types. *Earth's Future* **2023**, *11*, e2022EF003108. [CrossRef]
28. Xue, B.; Wang, G.; Xiao, J.; Helman, D.; Sun, W.; Wang, J.; Liu, T. Global convergence but regional disparity in the hydrological resilience of ecosystems and watersheds to drought. *J. Hydrol.* **2020**, *591*, 125589. [CrossRef]
29. Shao, X.; Zhang, Y.; Ma, N.; Zhang, X.; Tian, J.; Xu, Z.; Liu, C. Drought-induced ecosystem resistance and recovery observed at 118 flux tower stations across the globe. *Agric. For. Meteorol.* **2024**, *356*, 110170. [CrossRef]
30. Sheoran, S.; Kaur, Y.; Kumar, S.; Shukla, S.; Rakshit, S.; Kumar, R. Recent advances for drought stress tolerance in maize (*Zea mays* L.): Present status and future prospects. *Front. Plant Sci.* **2022**, *13*, 872566. [CrossRef]
31. Stuart-Haëntjens, E.; De Boeck, H.J.; Lemoine, N.P.; Mänd, P.; Kröel-Dulay, G.; Schmidt, I.K.; Jentsch, A.; Stampfli, A.; Anderegg, W.R.; Bahn, M. Mean annual precipitation predicts primary production resistance and resilience to extreme drought. *Sci. Total Environ.* **2018**, *636*, 360–366. [CrossRef]
32. DeSoto, L.; Cailleret, M.; Sterck, F.; Jansen, S.; Kramer, K.; Robert, E.M.; Aakala, T.; Amoroso, M.M.; Bigler, C.; Camarero, J.J. Low growth resilience to drought is related to future mortality risk in trees. *Nat. Commun.* **2020**, *11*, 545. [CrossRef]
33. Li, G.; Chen, W.; Li, R.; Zhang, X.; Liu, J. Assessing the spatiotemporal dynamics of ecosystem water use efficiency across China and the response to natural and human activities. *Ecol. Indic.* **2021**, *126*, 107680. [CrossRef]
34. Fernández-Guisuraga, J.M.; Calvo, L.; Suárez-Seoane, S. Comparison of pixel unmixing models in the evaluation of post-fire forest resilience based on temporal series of satellite imagery at moderate and very high spatial resolution. *ISPRS J. Photogramm. Remote Sens.* **2020**, *164*, 217–228. [CrossRef]
35. von Keyserlingk, J.; de Hoop, M.; Mayor, A.; Dekker, S.C.; Rietkerk, M.; Förster, S. Resilience of vegetation to drought: Studying the effect of grazing in a Mediterranean rangeland using satellite time series. *Remote Sens. Environ.* **2021**, *255*, 112270. [CrossRef]

36. Zhang, M.; Yang, L.; Hao, R.; Bai, X.; Wang, Y.; Yu, X. Drought-tolerant plant growth-promoting rhizobacteria isolated from jujube (*Ziziphus jujuba*) and their potential to enhance drought tolerance. *Plant Soil* **2020**, *452*, 423–440. [CrossRef]
37. Wang, C.; Zhang, J.; Li, J.; Yang, Y.; Zhang, Z.; Chai, Q.; Xie, J. Genome-wide identification and functional analysis of the trehalose-6-phosphate phosphatase (TPP) gene family in tomato (*Solanum lycopersicum* L.) and the role of SlTPP3 under NaCl stress. *Sci. Hortic.* **2025**, *345*, 114161. [CrossRef]
38. Ilyas, M.; Nisar, M.; Khan, N.; Hazrat, A.; Khan, A.H.; Hayat, K.; Fahad, S.; Khan, A.; Ullah, A. Drought tolerance strategies in plants: A mechanistic approach. *J. Plant Growth Regul.* **2021**, *40*, 926–944. [CrossRef]
39. Müller, L.M.; Bahn, M. Drought legacies and ecosystem responses to subsequent drought. *Glob. Change Biol.* **2022**, *28*, 5086–5103. [CrossRef]
40. Henkhaus, N.; Bartlett, M.; Gang, D.; Grumet, R.; Jordon-Thaden, I.; Lorence, A.; Lyons, E.; Miller, S.; Murray, S.; Nelson, A. Plant science decadal vision 2020–2030: Reimagining the potential of plants for a healthy and sustainable future. *Plant Direct* **2020**, *4*, e00252. [CrossRef] [PubMed]
41. Mcdowell, N.G.; Pockman, W.T.; Allen, C.D.; Breshears, D.D.; Cobb, N.; Kolb, T.; Plaut, J.; Sperry, J.; West, A.; Williams, D.G. Mechanisms of plant survival and mortality during drought: Why do some plants survive while others succumb to drought? *New Phytol.* **2008**, *178*, 719–739. [CrossRef]
42. Hartmann, H.; Trumbore, S. Understanding the roles of nonstructural carbohydrates in forest trees–from what we can measure to what we want to know. *New Phytol.* **2016**, *211*, 386–403. [CrossRef]
43. Blum, A. Osmotic adjustment is a prime drought stress adaptive engine in support of plant production. *Plant Cell Environ.* **2017**, *40*, 4–10. [CrossRef]
44. Mohammadi Alagoz, S.; Zahra, N.; Hajiaghaei Kamrani, M.; Asgari Lajayer, B.; Nobaharan, K.; Astatkie, T.; Siddique, K.H.; Farooq, M. Role of root hydraulics in plant drought tolerance. *J. Plant Growth Regul.* **2023**, *42*, 6228–6243. [CrossRef]
45. Hoover, D.L.; Pfennigwerth, A.A.; Duniway, M.C. Drought resistance and resilience: The role of soil moisture–plant interactions and legacies in a dryland ecosystem. *J. Ecol.* **2021**, *109*, 3280–3294. [CrossRef]
46. Jiang, H.; Chen, X.; Xu, G.; Chen, J.; Song, D.; Lv, M.; Guo, H.; Chen, J. Plant Adaptation and Soil Shear Strength: Unraveling the Drought Legacy in *Amorpha fruticosa*. *Plants* **2025**, *14*, 179. [CrossRef]
47. Ghadirnezhad Shiade, S.R.; Fathi, A.; Taghavi Ghasemkheili, F.; Amiri, E.; Pessarakli, M. Plants' responses under drought stress conditions: Effects of strategic management approaches—A review. *J. Plant Nutr.* **2023**, *46*, 2198–2230. [CrossRef]
48. Suzuki, N.; Koussevitzky, S.; Mittler, R.; Miller, G. ROS and redox signalling in the response of plants to abiotic stress. *Plant Cell Environ.* **2012**, *35*, 259–270. [CrossRef]
49. Tian, H.; De Smet, I.; Ding, Z. Shaping a root system: Regulating lateral versus primary root growth. *Trends Plant Sci.* **2014**, *19*, 426–431. [CrossRef] [PubMed]
50. Waseem, M.; Ali, A.; Tahir, M.; Nadeem, M.; Ayub, M.; Tanveer, A.; Ahmad, R.; Hussain, M. Mechanism of drought tolerance in plant and its management through different methods. *Cont. J. Agric. Sci.* **2011**, *5*, 10–25.
51. Li, X.; Bao, J.; Wang, J.; Blackman, C.; Tissue, D. Antecedent drought condition affects responses of plant physiology and growth to drought and post-drought recovery. *Front. For. Glob. Change* **2021**, *4*, 704470. [CrossRef]
52. Zhao, Z.; Cai, Y.; Fu, M.; Bai, Z. Response of the soils of different land use types to drought: Eco-physiological characteristics of plants grown on the soils by pot experiment. *Ecol. Eng.* **2008**, *34*, 215–222. [CrossRef]
53. Gupta, A.; Rico-Medina, A.; Caño-Delgado, A.I. The physiology of plant responses to drought. *Science* **2020**, *368*, 266–269. [CrossRef] [PubMed]
54. Lee, S.-H.; Li, C.-W.; Koh, K.W.; Chuang, H.-Y.; Chen, Y.-R.; Lin, C.-S.; Chan, M.-T. MSRB7 reverses oxidation of GSTF2/3 to confer tolerance of *Arabidopsis thaliana* to oxidative stress. *J. Exp. Bot.* **2014**, *65*, 5049–5062. [CrossRef]
55. Batool, A.; Yue, D.-X.; Xiao, Y.-L.; Li, S.-S.; Duan, H.-X.; Haq, Z.; Ahmed, K.; Zhao, L.; Zhu, L.; Xiong, Y. Plant tolerance to drought stress: Complexity and mechanism across physiological, molecular and biochemical scales. *Int. J. Appl. Exp. Biol.* **2024**, *3*, 159–175. [CrossRef]
56. Yao, Y.; Fu, B.; Liu, Y.; Li, Y.; Wang, S.; Zhan, T.; Wang, Y.; Gao, D. Evaluation of ecosystem resilience to drought based on drought intensity and recovery time. *Agric. For. Meteorol.* **2022**, *314*, 108809. [CrossRef]
57. Zou, J.; Ding, J.; Huang, S.; Liu, B. Ecosystem resistance and resilience after dry and wet events across central Asia based on remote sensing data. *Remote Sens.* **2023**, *15*, 3165. [CrossRef]
58. Kang, J.; Shen, H.; Liu, Y.; Ma, P.; Wu, B.; Xu, L.; Fang, J. Drought dimensions impact birch resistance and resilience and their determining factors across semiarid forests of northern China. *Agric. For. Meteorol.* **2025**, *360*, 110314. [CrossRef]
59. Kalra, A.; Goel, S.; Elias, A.A. Understanding role of roots in plant response to drought: Way forward to climate-resilient crops. *Plant Genome* **2024**, *17*, e20395. [CrossRef] [PubMed]
60. Singh, N.; Giri, M.K.; Chattopadhyay, D. Lighting the path: How light signaling regulates stomatal movement and plant immunity. *J. Exp. Bot.* **2025**, *76*, 769–786. [CrossRef]

61. Kumar, V.; Sharma, K.V.; Pham, Q.B.; Srivastava, A.K.; Bogireddy, C.; Yadav, S. Advancements in drought using remote sensing: Assessing progress, overcoming challenges, and exploring future opportunities. *Theor. Appl. Climatol.* **2024**, *155*, 4251–4288. [CrossRef]
62. Shoaib, M.; Banerjee, B.P.; Hayden, M.; Kant, S. Roots' drought adaptive traits in crop improvement. *Plants* **2022**, *11*, 2256. [CrossRef]
63. Van Ruijven, J.; Berendse, F. Diversity enhances community recovery, but not resistance, after drought. *J. Ecol.* **2010**, *98*, 81–86. [CrossRef]
64. Gao, C.; Xu, L.; Montoya, L.; Madera, M.; Hollingsworth, J.; Chen, L.; Purdom, E.; Singan, V.; Vogel, J.; Hutmacher, R.B. Co-occurrence networks reveal more complexity than community composition in resistance and resilience of microbial communities. *Nat. Commun.* **2022**, *13*, 3867. [CrossRef]
65. Köpp Hollunder, R.; Garbin, M.L.; Rubio Scarano, F.; Mariotte, P. Regional and local determinants of drought resilience in tropical forests. *Ecol. Evol.* **2022**, *12*, e8943. [CrossRef] [PubMed]
66. Zvanarou, S.; Vágnerová, R.; Mackievic, V.; Usnich, S.; Smolich, I.; Sokolik, A.; Yu, M.; Huang, X.; Angelis, K.; Demidchik, V. Salt stress triggers generation of oxygen free radicals and DNA breaks in *Physcomitrella patens* protonema. *Environ. Exp. Bot.* **2020**, *180*, 104236. [CrossRef]
67. Kumar, S.; Sachdeva, S.; Bhat, K.; Vats, S. Plant responses to drought stress: Physiological, biochemical and molecular basis. In *Biotic and Abiotic Stress Tolerance in Plants*; Springer: Singapore, 2018; pp. 1–25.
68. Talbi, S.; Rojas, J.A.; Sahrawy, M.; Rodríguez-Serrano, M.; Cárdenas, K.E.; Debouba, M.; Sandalio, L.M. Effect of drought on growth, photosynthesis and total antioxidant capacity of the saharan plant *Oudeneya africana*. *Environ. Exp. Bot.* **2020**, *176*, 104099. [CrossRef]
69. Huang, Q.-L.; Zhang, M.-M.; Li, C.-S.; Li, B.-Y.; Zhuo, S.-L.; Yang, Y.-S.; Chen, Y.-D.; Zhong, A.-N.; Liu, H.-Y.; Lai, W.-F. Response mechanism of water status and photosynthetic characteristics of *Cotoneaster multiflorus* under drought stress and rehydrated conditions. *Front. Plant Sci.* **2025**, *15*, 1457955. [CrossRef]
70. Liu, S.; Liu, H.; Jiao, J.; Yin, J.; Lu, T.J.; Xu, F. Biomechanics in plant resistance to drought. *Acta Mech. Sin.* **2020**, *36*, 1142–1157. [CrossRef]
71. Barbut, F.R.; Cavel, E.; Donev, E.N.; Gaboreanu, I.; Urbancsok, J.; Pandey, G.; Demailly, H.; Jiao, D.; Yassin, Z.; Derba-Maceluch, M. Integrity of xylan backbone affects plant responses to drought. *Front. Plant Sci.* **2024**, *15*, 1422701. [CrossRef]
72. Anderegg, W.R.; Konings, A.G.; Trugman, A.T.; Yu, K.; Bowling, D.R.; Gabbitas, R.; Karp, D.S.; Pacala, S.; Sperry, J.S.; Sulman, B.N. Hydraulic diversity of forests regulates ecosystem resilience during drought. *Nature* **2018**, *561*, 538–541. [CrossRef] [PubMed]
73. Sinclair, G.C.; Travadon, R.; Eschen, P.J.; Wallis, C.; Baumgartner, K.; Delmas, C.E.; Hnizdor, J.F.; Bartlett, M.K. Differential physiological responses of resistant and susceptible grape cultivars to *Eutypa* dieback. *J. Exp. Bot.* **2025**, *76*, 3172–3185. [CrossRef]
74. Palma-Bautista, C.; Torra, J.; Garcia, M.J.; Bracamonte, E.; Rojano-Delgado, A.M.; Alcántara-de la Cruz, R.; De Prado, R. Reduced absorption and impaired translocation endows glyphosate resistance in *Amaranthus palmeri* harvested in glyphosate-resistant soybean from Argentina. *J. Agric. Food Chem.* **2019**, *67*, 1052–1060. [CrossRef]
75. Li, X.; Jia, W.; Zheng, J.; Ma, L.; Duan, Q.; Du, W.; Cui, G.; Wang, X.; Wang, J. Application of morphological and physiological markers for study of drought tolerance in *Lilium* varieties. *Horticulturae* **2022**, *8*, 786. [CrossRef]
76. Ingrisch, J.; Umlauf, N.; Bahn, M. Functional thresholds alter the relationship of plant resistance and recovery to drought. *Ecology* **2023**, *104*, e3907. [CrossRef] [PubMed]
77. Kesahvarznia, R.; Shahbazi, M.; Mohammadi, V.; Hosseini Salekdeh, G.; Ahmadi, A.; Mohseni-Fard, E. The impact of barley root structure and physiological traits on drought response. *Iran. J. Field Crop Sci.* **2014**, *45*, 553–563.
78. Willis, K.J.; Jeffers, E.S.; Tovar, C. What makes a terrestrial ecosystem resilient? *Science* **2018**, *359*, 988–989. [CrossRef] [PubMed]
79. Liu, D.; Wang, T.; Peñuelas, J.; Piao, S. Drought resistance enhanced by tree species diversity in global forests. *Nat. Geosci.* **2022**, *15*, 800–804. [CrossRef]
80. Backhaus, L.; Albert, G.; Cuchietti, A.; Jaimes Nino, L.M.; Fahs, N.; Lisner, A.; Kolář, V.; Kermavnar, J.; Widmer, S.; Zimmermann, Z. Shift from trait convergence to divergence along old-field succession. *J. Veg. Sci.* **2021**, *32*, e12986. [CrossRef]
81. Roces-Díaz, J.V.; García-Valdés, R.; de Cáceres, M.; Descals, A.; Hurtado, P.; Lloret, F.; Espelta, J.M.; Álvarez-Martínez, J.M.; Batllori, E.; Martínez-Vilalta, J. Water availability, stand structure, and hydraulic trait diversity drive forest stability. *For. Ecol. Manag.* **2025**, *594*, 122946. [CrossRef]
82. Arthur, C.M.; Dech, J.P. Species composition determines resistance to drought in dry forests of the Great Lakes–St. Lawrence forest region of central Ontario. *J. Veg. Sci.* **2016**, *27*, 914–925. [CrossRef]
83. García-Valdés, R.; Vayreda, J.; Retana, J.; Martínez-Vilalta, J. Low forest productivity associated with increasing drought-tolerant species is compensated by an increase in drought-tolerance richness. *Glob. Change Biol.* **2021**, *27*, 2113–2127. [CrossRef]
84. Komainda, M.; Küchenmeister, F.; Küchenmeister, K.; Kayser, M.; Wrage-Mönnig, N.; Isselstein, J. Drought tolerance is determined by species identity and functional group diversity rather than by species diversity within multi-species swards. *Eur. J. Agron.* **2020**, *119*, 126116. [CrossRef]

85. Wang, X.-Y.; Ge, Y.; Gao, S.; Chen, T.; Wang, J.; Yu, F.-H. Evenness alters the positive effect of species richness on community drought resistance via changing complementarity. *Ecol. Indic.* **2021**, *133*, 108464. [CrossRef]
86. Ding, J.; Buotte, P.; Bales, R.; Christoffersen, B.; Fisher, R.A.; Goulden, M.; Knox, R.; Kueppers, L.; Shuman, J.; Xu, C. Coordination of rooting, xylem, and stomatal strategies explains the response of conifer forest stands to multi-year drought in the southern Sierra Nevada of California. *Biogeosciences* **2023**, *20*, 4491–4510. [CrossRef]
87. Haghpanah, M.; Hashemipetroudi, S.; Arzani, A.; Araniti, F. Drought tolerance in plants: Physiological and molecular responses. *Plants* **2024**, *13*, 2962. [CrossRef]
88. Castagneri, D.; Vacchiano, G.; Hacket-Pain, A. Meta-analysis reveals different competition effects on tree growth resistance and resilience to drought. *Ecosystems* **2022**, *25*, 30–43. [CrossRef]
89. Bai, Y.; Tang, Z. Enhanced effects of species richness on resistance and resilience of global tree growth to prolonged drought. *Proc. Natl. Acad. Sci. USA* **2024**, *121*, e2410467121. [CrossRef]
90. Bartlett, M.K.; Zhang, Y.; Yang, J.; Kreidler, N.; Sun, S.W.; Lin, L.; Hu, Y.H.; Cao, K.F.; Sack, L. Drought tolerance as a driver of tropical forest assembly: Resolving spatial signatures for multiple processes. *Ecology* **2016**, *97*, 503–514. [CrossRef] [PubMed]
91. O'Brien, M.J.; Reynolds, G.; Ong, R.; Hector, A. Resistance of tropical seedlings to drought is mediated by neighbourhood diversity. *Nat. Ecol. Evol.* **2017**, *1*, 1643–1648. [CrossRef]
92. Wahab, A.; Muhammad, M.; Munir, A.; Abdi, G.; Zaman, W.; Ayaz, A.; Khizar, C.; Reddy, S.P.P. Role of arbuscular mycorrhizal fungi in regulating growth, enhancing productivity, and potentially influencing ecosystems under abiotic and biotic stresses. *Plants* **2023**, *12*, 3102. [CrossRef]
93. Xu, Q.; Yu, R.; Guo, L. Evaluation of forest ecosystem resilience to drought considering lagged effects of drought. *Ecol. Evol.* **2024**, *14*, e70281. [CrossRef] [PubMed]

MDPI AG
Grosspeteranlage 5
4052 Basel
Switzerland
Tel.: +41 61 683 77 34

Water Editorial Office
E-mail: water@mdpi.com
www.mdpi.com/journal/water

www.ingramcontent.com/pod-product-compliance
Lightning Source LLC
Chambersburg PA
CBHW061617210926
51850CB00009B/2351
* 9 7 8 3 7 2 5 8 6 6 5 2 6 *